Plumbing without a Plumber

Plumbing without a Plumber

by PETER JONES

Butterick Publishing

Book Design by Jos. Trautwein

Illustrations by Ralph Haarup

Library of Congress Cataloging in Publication Data
Jones, Peter, 1934–
Plumbing without a plumber.
Includes Index.
1. Plumbing—Amateurs' manuals. I. Title.
TH6124.J67 696'.1 79–26583
ISBN 0-88421-084-7

Manufactured and printed in the United States of America.
Published simultaneously in the USA and Canada.

Contents

INTRODUCTION

How to Use This Book

MORE THAN 90% of all plumbing repairs in the home can be made by practically anyone. In fact, assembling an entire plumbing system is really not very difficult. But because plumbing repairs and installations require special tools, homeowners have traditionally hired professionals to get the job done—at costs that are often outrageous. Fortunately, that tradition is changing. The price of metal pipe has risen with the cost of labor to awesome levels, and as a result homeowners and apartment dwellers are learning to make their own plumbing repairs and improvements. And they're saving money.

This book is designed to help you understand your plumbing system and the work you can do yourself to improve it. Dozens of repairs and installations are detailed in step-by-step procedures and simple, how-to illustrations. In addition, you will find handy what-to-do charts at the beginning of the chapters covering faucet and toilet repairs; these provide a listing of every possible malfunction and how to fix it, along with page references to in-depth information and specific repair procedures.

In general, the first part of the book details major and minor plumbing repairs, in the order they most frequently occur. The second part outlines procedures for improving your home plumbing system, which may involve working with different pipes, adding new plumbing lines, or installing or renovating a bathroom. Between the covers of this book you will find everything you need to know to do practically any kind of plumbing in your home. Remember that the information has been organized according to how often you will need it: the procedure for unclogging drains, for instance, will be found well before the instructions for installing a new bathtub.

A few years ago, at about the same time homeowners began to teach themselves how to cut 4″ cast iron pipe and started buying propane torches to practice making sweat-soldered joints in copper tubing, along came plastic pipe. Plastic pipe and tubing is a revolution in itself, and it's the wave of the future in plumbing that you can work with now—quickly and inexpensively. Plastic pipe enables anyone to assemble an entire home plumbing system without the services of a professional. All you need is a hacksaw, a can of special solvent-weld glue, some common sense, and an idea of what you want the pipes to do.

While the steps of each repair procedure in this book offer the most precise information possible, you may encounter a unique situation or run across a piece of equipment that appeared on the market after the book was published (the new plastic faucets, for instance, are similar to the metal ones in design, but are assembled differently). If so, view the repair procedure in its general context, and remember that the basic principles of the approach will not change.

Although there is nothing very complicated about making repairs to your plumbing system, you must always follow one cardinal rule: *Every pipe connection you make must be absolutely watertight.*

The basic equipment for plumbing repairs includes a box of assorted faucet washers and a roll of plastic pipe-sealing tape. Each costs less than a dollar. You should also have the following tools on hand:

1 standard-blade screwdriver
1 phillips-head screwdriver
1 10" adjustable-end (open-end) wrench
1 10" or 12" pipe wrench
1 rubber force cup (plunger)

You could, of course, accumulate a large number of specialty tools, but don't spend the money until you need them. Chances are you never will.

CHAPTER 1

Coping with Plumbing Emergencies

AN EMERGENCY in your home plumbing system has only one definition: Water is pouring out of the system. It may be flooding through a wall, dripping from the ceiling, spurting from the joint between a pipe and its fitting, or gushing through a split in the pipe itself.

Shutoff Valves

The first step in coping with a plumbing emergency is to stop the water flow by closing the **main shutoff valve.** The valve is located at the water supply service entrance, which is a ¾″- or 1″-wide pipe that enters your house from the street. The entrance is most likely in your basement, and as soon as the pipe enters the house it passes through a valve. Main shutoff valves are usually globe or gate valves that have a wheel on top. Turning the wheel clockwise will drive the long plunger to the bottom of the unit and stop the flow of water entering the house. You may discover that it takes a great deal of turning before the wheel stops moving. Don't get discouraged; unless the valve is broken it will eventually come to a halt, at which point water will no

Hang a tag on the main shutoff valve so that everyone in your household can find it in case of an emergency.

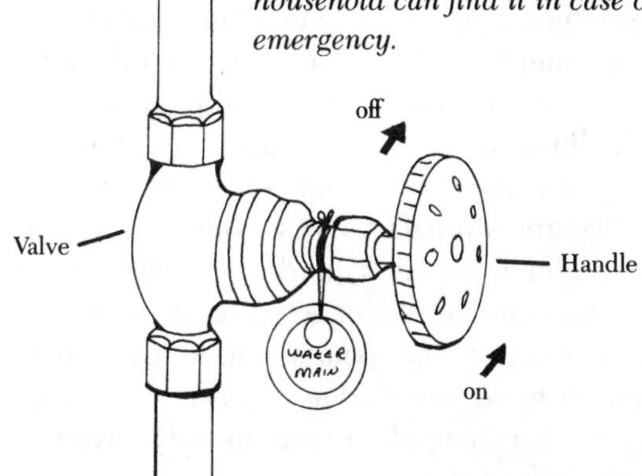

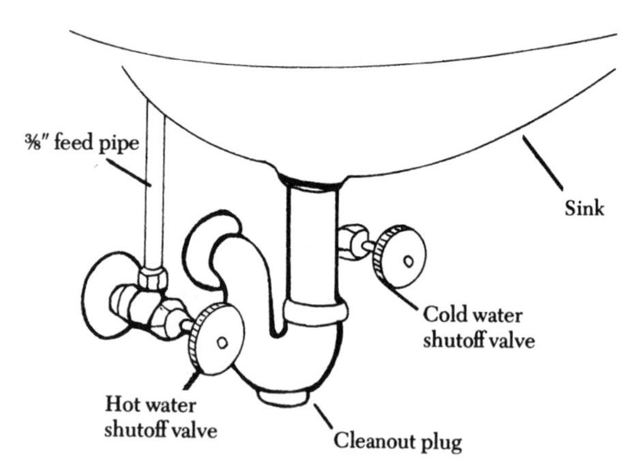

Most modern plumbing systems have shutoff valves near each fixture.

longer be entering your plumbing system. The emergency part of your crisis is over; now you must locate and solve the problem within the system.

As soon as the main shutoff valve is closed, open all of the faucets in your bathrooms, kitchen, and laundry room to allow any water in the pipes to drain off. The water pressure in the supply pipes is between 40 and 60 pounds per square inch (p.s.i.), which is why the water is able to gush out of your faucet spouts every time you open a tap. By opening the faucets you will quickly drain the pipes, and with the main valve closed no more water can leak from the system to damage your walls, ceilings, or floors.

An alternative to closing the main shutoff valve, which may be quicker than wading into the basement, is to close the shutoff valve nearest the disaster area. Usually, the water supply lines have shutoff valves near every fixture, or as soon as the pipe emerges from the wall. **Fixture shutoff valves**, then, should be found under every sink, under the left side of every toilet, and on the pipes hanging from your cellar ceiling at the points where the pipes branch off to service various rooms. Be aware, however, that some older systems only have the main shutoff valve.

Fixture shutoff valves enable you to close off individual fixtures from the water supply without interrupting service to the rest of the house. They function in the same manner as a main shutoff valve, or any faucet for that matter: a handle is turned clockwise to close off the water to the fixture, and counterclockwise to open it. If fixture shutoff valves do not exist at the time you are making a plumbing repair, it is a good idea to add them, if only for your own convenience in the future.

Stuck or Damaged Valves

As a rule, most of the shutoff valves in your home are permanently open. They may remain open for years, even decades, without any need to be closed. All of this inactivity may cause them to leak when you finally do try to close them. Most often, the leak will occur around the packing nut, which is at the base of the handle stem. The nut is filled with a putty-like packing or may rest atop a washer, which can disintegrate simply from years of disuse. If this has occurred, use a pipe wrench to tighten the nut a quarter turn, or until the leak stops. If that fails, close off the water supply leading to the valve and repair it as you would any faucet.

The one critical emergency that can occur to valves involves the main shutoff valve. If you have a plumbing crisis somewhere in the house and discover that the main valve will not shut off, you have a definite problem. The only other way that water can be kept from entering your system is to close it off at the street, and this can only be done by a municipal agency (usually the water supply department). Municipal water supply departments normally have standby crews available around the clock: if you call to explain your emergency and ask them to shut off your water supply, they will usually respond to your request quickly. Very often (but not always) the municipality will take responsibility for either repairing or replacing the main shutoff valve and then bill you. Otherwise, the repair is not a difficult one for you to make by yourself (see pages 35–36 for repair and replacement of valves).

Emergency Repair of Pipes

With the water supply shut off, the following interim repairs can be made quickly to get your plumbing system working properly again. But in no case other than with unfreezing pipes should any such "patchwork" be considered permanent. At some later date when you have the time, equipment, and materials, replace the interim repair with a proper, permanent renovation.

Frozen Pipes

There are several ways of unfreezing water supply pipes that have become clogged with ice. No matter which approach you use, bear in mind that the pipe is cold: you will have to be patient and work slowly, or you may split the metal.

Heating cables—If you own a heating cable, wrap it around the frozen pipe and plug it in. Open the nearest faucet attached to the pipe to allow the water to flow as it melts.

Insulation—Pipe-insulating tape or strips of standard house insulation can also be wrapped around a frozen pipe. The insulation will take longer to warm the pipe than a heating cable, but eventually it will do the job. Again, leave the nearest faucet(s) open to allow the water to flow as it melts.

Hot rags—The time-honored method of unfreezing pipes is to wrap them in rags soaked in hot water:

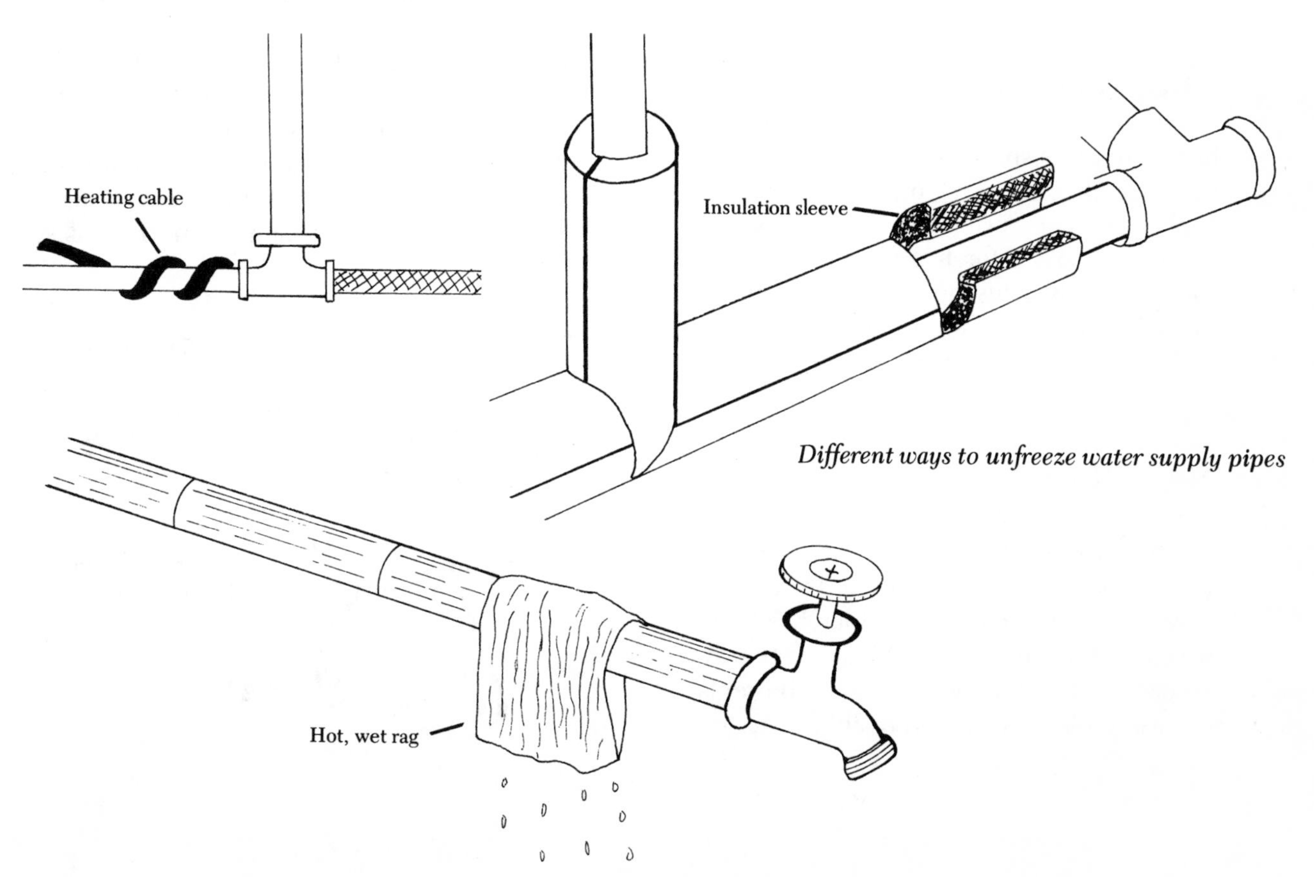

Different ways to unfreeze water supply pipes

1. Open the faucets nearest the frozen area.
2. Dip rags in boiling water and wring them out.
3. Wrap the rags around the pipe, beginning nearest the open faucet(s) and working back toward the main shutoff valve.
4. Reheat the rags frequently and reapply them to the pipe.

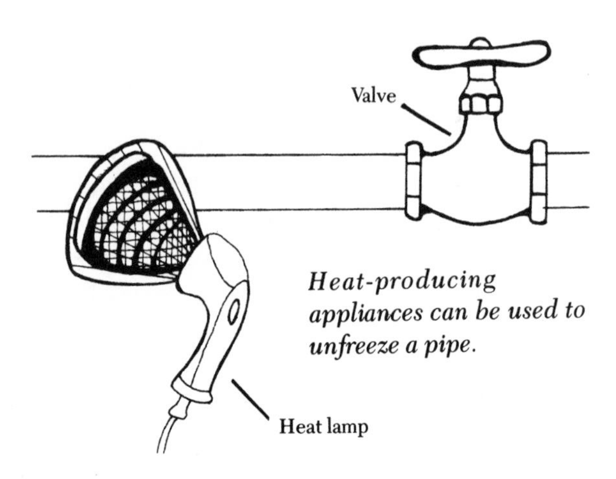

Heat-producing appliances can be used to unfreeze a pipe.

Hair blowers/dryers, space heaters, torches—You can try to warm the pipes with a hair blower/dryer, a space heater, even a propane or blow torch, or any other heat-producing appliance. When using any of these appliances, and particularly a torch, keep the heat moving back and forth along the length of the pipe until water flow is restored in the pipe. Do not concentrate on just one part of the pipe.

Split Pipes

A pipe can split from a number of causes: it may have frozen, or been hit by something heavy, or simply corroded under the constant pressure of moving water. Whatever the cause, the damaged section of pipe should be replaced with a new piece as soon as possible. As a stopgap repair (*providing the damage is no more than 4" in length*), you can use a dresser coupling, a compression clamp, or a makeshift clamp of your own invention.

Dresser couplings—These can be used with a minimum of effort. The coupling must be designed to fit around the pipe that has split; the most common pipe diameters are ¾", ½", and ⅜".

1. Shut off the valve nearest the split.
2. Open the faucets nearest the split pipe and drain as much water as you can from the pipe.
3. Using a hacksaw, cut the pipe at about an inch beyond both ends of the split. You cannot remove more than 6" of pipe or the coupling will be unable to span the gap.
4. Remove the compression collars at each end of the coupling body and slide them onto the severed ends of the pipe.
5. Push one end of the coupling body over one of the pipe ends, then connect the opposite end.
6. Slide the pipe clamps over the ends of the coupling body and tighten them with a wrench until they are snug.
7. Open the shutoff valve. Check both connections for leakage. If there is any sign of water, tighten the clamps until the leakage stops.

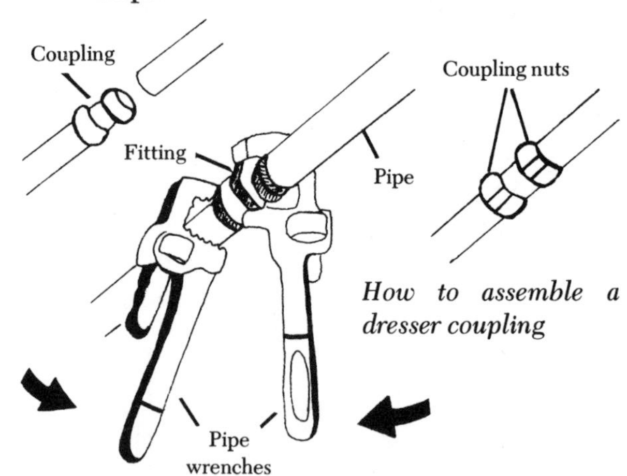

How to assemble a dresser coupling

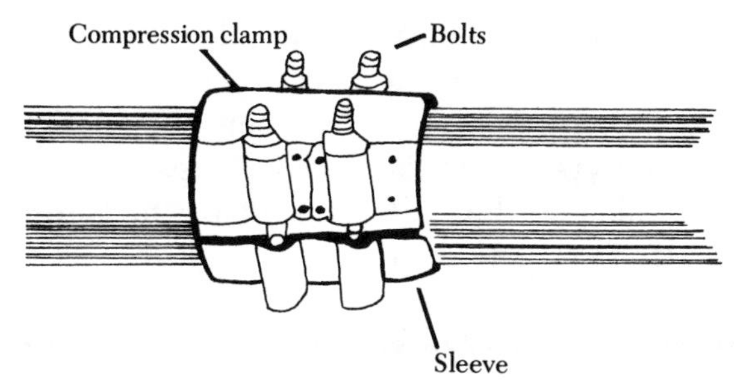

Bolted compression clamp

Compression clamps—Compression clamps consist of two curved pieces of metal which are hinged together to fit around a neoprene sleeve. The clamp is held shut with bolts or screws. The clamp must be the proper diameter for the pipe being repaired.

1. Close the valve controlling the damaged pipe and open the nearest faucets to drain all water.
2. Open the clamp and place it around the split in the pipe. The clamp should extend approximately one inch beyond both ends of the damaged area.
3. Place the bolts in their holes in the clamp and tighten them alternately, so that the clamp is closed evenly.
4. Open the water valve and inspect the clamp for leakage. Tighten the bolts until all signs of water disappear.

You may be able to effect a repair using a pad of rubber or heavy plastic and a hose clamp. Wrap the split with the pad and fasten the hose clamp around it.

Another possibility is to cut a tin can along its seam and wrap it around a rubber or plastic pad encircling the split. Hold the can together with blocks of wood and a C-clamp.

Minor Pipe Damage

Because they are always filled with water and pressure, water supply pipes seldom have any minor damage; they either spring a leak or they remain watertight. But the drainpipes that lead from the bottom of each fixture to the house drain line (and eventually to the outside sewer) can develop minor cracks or pinholes, since they are usually empty and rarely under intense pressure. Such minor damage can be temporarily repaired, but the pipe should be replaced as soon as it is convenient to do so.

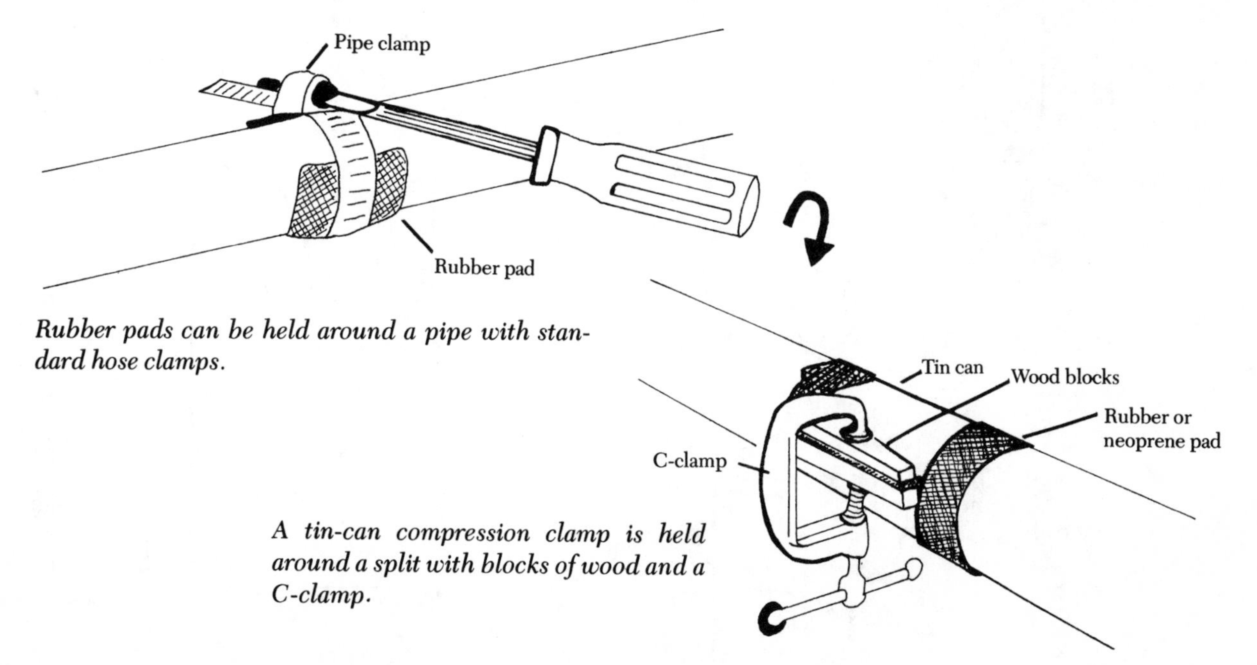

Rubber pads can be held around a pipe with standard hose clamps.

A tin-can compression clamp is held around a split with blocks of wood and a C-clamp.

REPAIRS WITH BEESWAX

Beeswax is available at many hardware and plumbing supply stores. It is used to fill pinholes and small cracks in pipe in this manner:

1. Heat the beeswax in a pan until it becomes pliable.
2. Push the wax into the damaged area using your fingers or a putty knife. Build the wax up over the damage.
3. Allow the wax to cool and harden.
4. When the wax has hardened, wrap several layers of friction tape around the repair.

REPAIRS WITH EPOXY

Epoxy can now be purchased in yellow and blue strips of plasticene that must be kneaded together until they blend into a uniform green color. The only danger with using epoxy is that once the two strips have been mixed, they must be used within about half an hour or the epoxy will not adhere to the metal pipe.

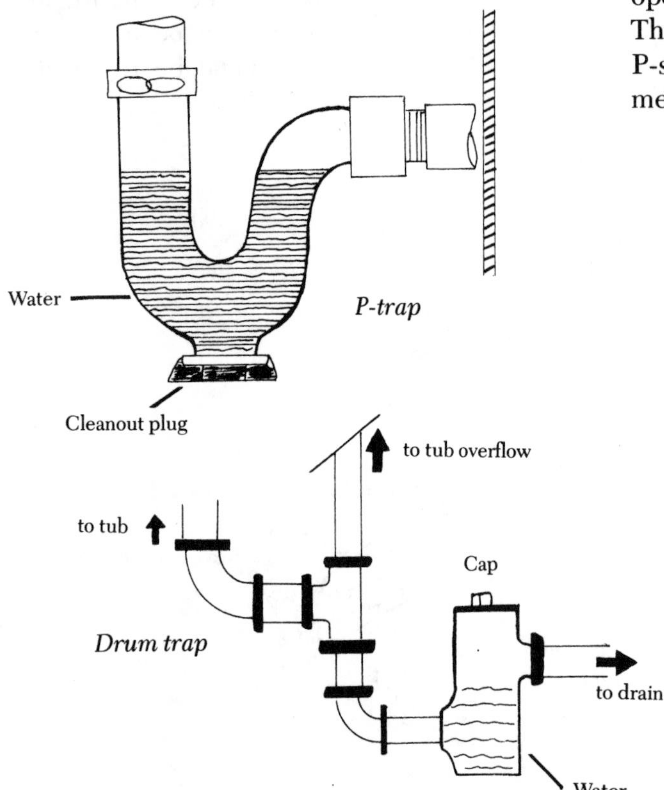

1. Knead the epoxy resin and its hardener until they become a uniform color.
2. Using a putty knife or your fingers, push the epoxy into the damaged area.
3. Allow 12 hours for the epoxy to harden before using the drain.

Unclogging Drains and Traps

Although not necessarily an emergency, clogged drains in sinks, bathtubs, and toilets are perhaps the most common malfunction to occur in home plumbing systems. When waste water stops flowing out of the drain in the bottom of a sink, you have several ways of clearing the blockage, beginning with a rubber force cup, or plunger. If this fails, the next procedure is to open the fixture trap and clean it of all debris. Theoretically, every sink and tub must have a P-shaped or S-shaped trap positioned immediately under the fixture (sometimes the trap

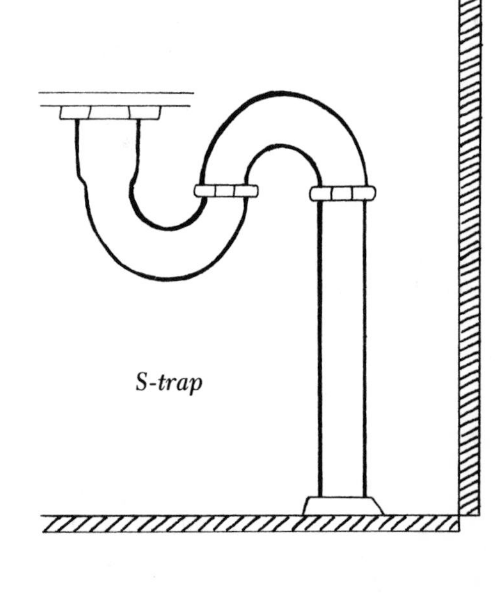

is below the floor under first-floor fixtures). Older bathtubs often have a drum trap, which is a metal canister that can be opened by rotating the screw-on cap that forms its top. A trap may or may not have a cleanout plug at the bottom of its curve; if it does not have one, it is connected to the drainpipe with a compression coupling that can be unscrewed to remove the trap.

If cleaning the trap does not free the drain, you can try using a compressed-air cleaner, an auger, or, as a last resort, a chemical drain opener.

PLUNGERS AND HOW TO USE THEM

Rubber force cups, more commonly known as **plungers** or the "plumber's friend," are sold in several designs. The most versatile of these has a retractable bulb that can be folded back to fit tightly over any drain opening, including the opening in toilet bowls. Plungers are most efficient when there are two or three inches of warm water in the bottom of the fixture; warm water helps dissolve grease in the drain, and provides a seal around the base of the plunger that improves its suction.

Before you use a plunger, remove any debris that has settled over the drain. If you are unclogging a sink or bathtub, stuff a damp rag in the fixture's overflow port to maximize pressure from the plunger.

1. Remove or open the drain stopper.
2. Place the plunger cup over the drain. Tilt it slightly to fill the cup with water.
3. With the cup squarely over the drain, push down on the handle, then pull the handle up. Repeat 4 or 5 times.
4. On the final upward stroke, yank the plunger off the drain. You may get a geyser of water, particularly if the blockage has been completely freed.
5. Continue plunging the drain until the clogging is removed, or until it becomes apparent that you will have to try some other approach.

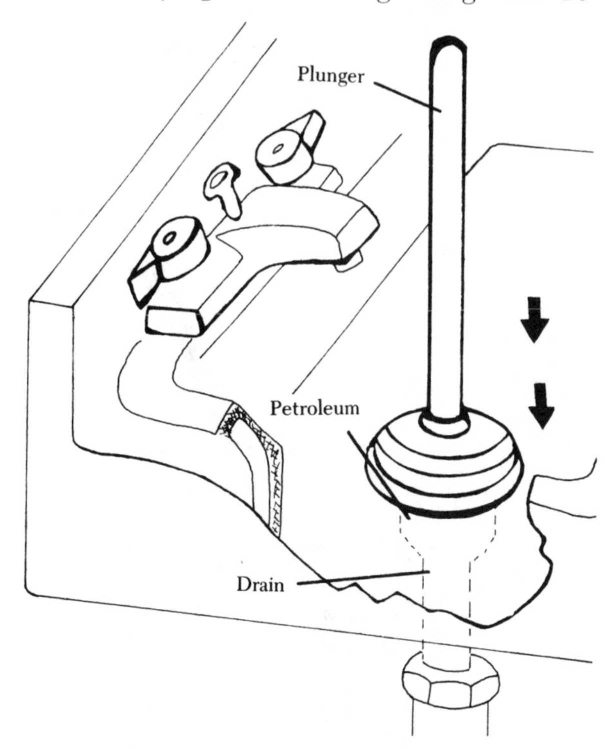

The seal around the rim of a plunger can be improved by coating the bottom rim of the cup with petroleum jelly.

TOILET AUGERS AND HOW TO USE THEM

Toilet augers, often called **plumbing snakes,** are sold in many different forms. Basically, an auger is either a long length of flat wire or a coil spring with a handle attached to one end and a wire bulb to the other. Some augers have a coil spring and an offset handle that can be rotated to make the spring act in much the same way as a brace and bit. Toilet augers are designed to be pushed up into the drainage port in toilet bowls:

1. Push the end of the tool into the drain and continue feeding it into the pipe until it comes to a halt, presumably against the obstruction in the pipe. The flexibility of

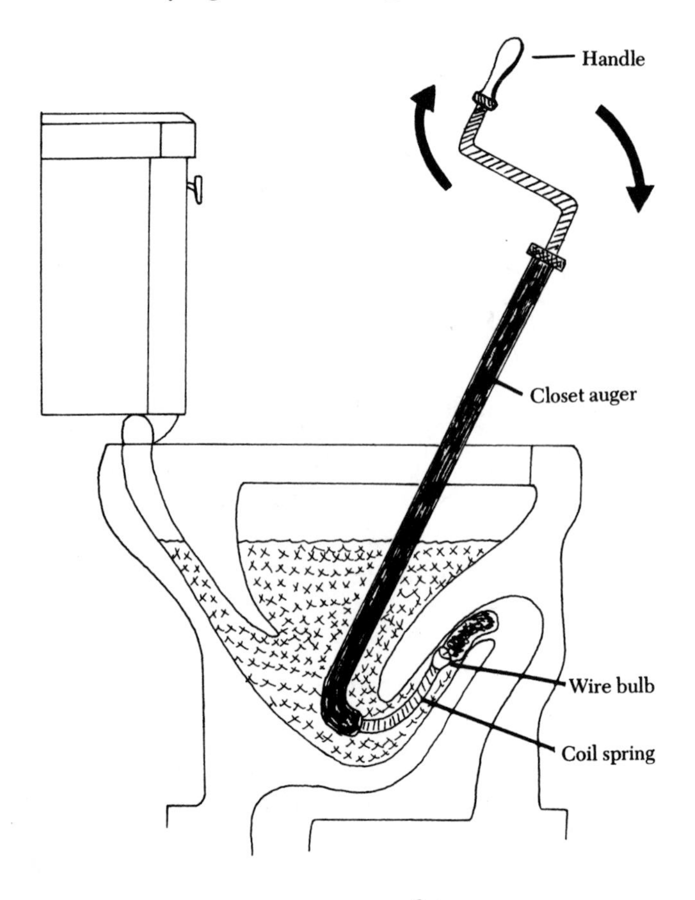

How to use a toilet auger

the cable is such that it can follow the bends and turns in the pipe without any difficulty.

2. If you are using the coiled-spring type of auger, turn its handle until you have bored through the clog. Use a ribbon-tape auger as a battering ram to pound against the obstruction until it is broken up.
3. When the obstruction is freed, flush the drain with water for a few minutes to clean away all remaining debris.

COMPRESSED-AIR CLEANERS

Compressed-air cleaners look like bicycle pumps without an air hose. At the bottom of the tool there is a tapered rubber grommet; this is inserted in the drain to form an airtight seal and enable the tool to exert tremendous pressure against any obstruction in any drain. There is a danger, however: if your plumbing system is more than 25 years old, it is quite possible for a compressed-air cleaner to blow the whole system apart at its seams. *Do not use a compressed-air cleaner on any older plumbing system.*

1. Remove the drain stopper.
2. Fill the bottom of the sink with about an inch of water.
3. Push the rubber end of the air cleaner into the drain.
4. Push down on the tool's plunger repeatedly until the clog is removed.

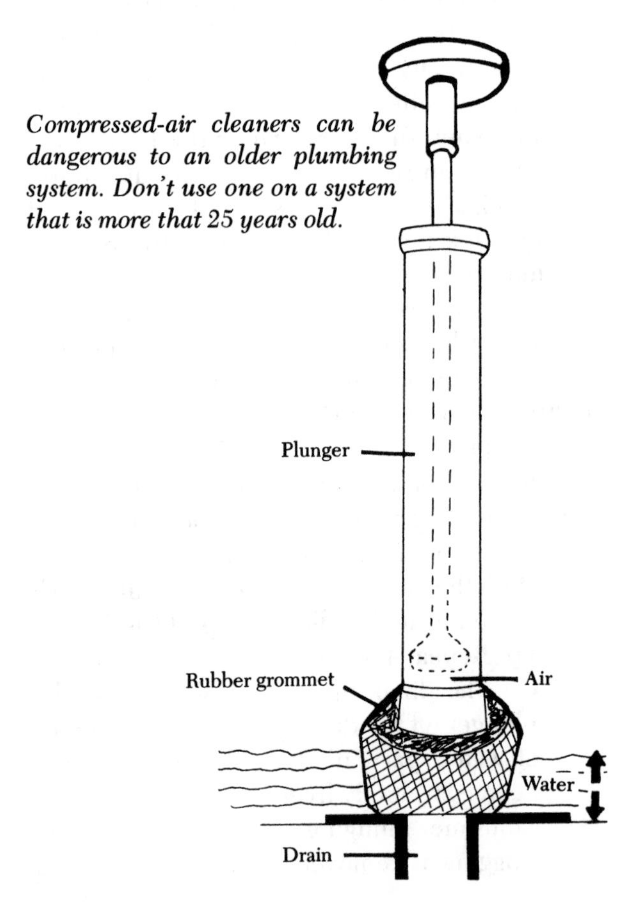

Compressed-air cleaners can be dangerous to an older plumbing system. Don't use one on a system that is more that 25 years old.

CHEMICAL DRAIN OPENERS

You can buy an assortment of chemical drain openers at almost any hardware store. All of them are relatively new products, with the exception of lye. Professional plumbers have been using lye for years to clear stuffed drains. In many instances it works extremely well, but be sure to follow the directions that are printed on the side of the can. You have to be careful when handling lye: it can be highly dangerous to your health.

Other chemical drain openers contain a dangerous amount of acid and often emit toxic fumes. Furthermore, they can be hazardous to your drainpipes. Rarely will any professional plumber use or recommend any of them. If you decide to buy one, be extremely careful to follow the manufacturer's instructions closely, both for your own sake and the well-being of your plumbing system.

Cleaning P-shaped and S-shaped Traps

The purpose of every trap is to hold water in the bottom of its curve, thus preventing sewer gases and insects from entering the house through the drains in your fixtures. The curve in the trap is also sharp enough to catch a considerable amount of debris, such as hair, which might otherwise become lodged elsewhere in the drain system. Should a plunger fail to unclog your drain, you can try to run an auger down the drain line. The curve in the trap may prevent the auger from functioning efficiently, though, or it may be that the trap itself is clogged. Do not hesitate to dismantle the trap and clean it. It can be easily removed from the fixture drain:

1. If the trap has a cleanout plug at the bottom of its curve, place a pan under the plug. Grip the cleanout plug with an open-end or pipe wrench and rotate it counterclockwise until it is unseated. The water in the bottom of the trap will run into the pan.

   If the trap has no cleanout plug, position a pan under the trap. There is a large nut at each end of the curve. Undo the nuts by turning them counterclockwise. The coupling nuts can often be undone by hand, but you may need to use a pipe wrench if the trap has not been removed in some time. When the coupling nuts are free, the trap can be pulled from the drain line. Pour the water in the trap into the pan.
2. If the trap has a cleanout plug, you can reach into it with your finger or a bent coat hanger to clean out any debris. If you have removed the trap, you can clean it with a piece of wire or any other tool that will fit inside it and bend around the curve.

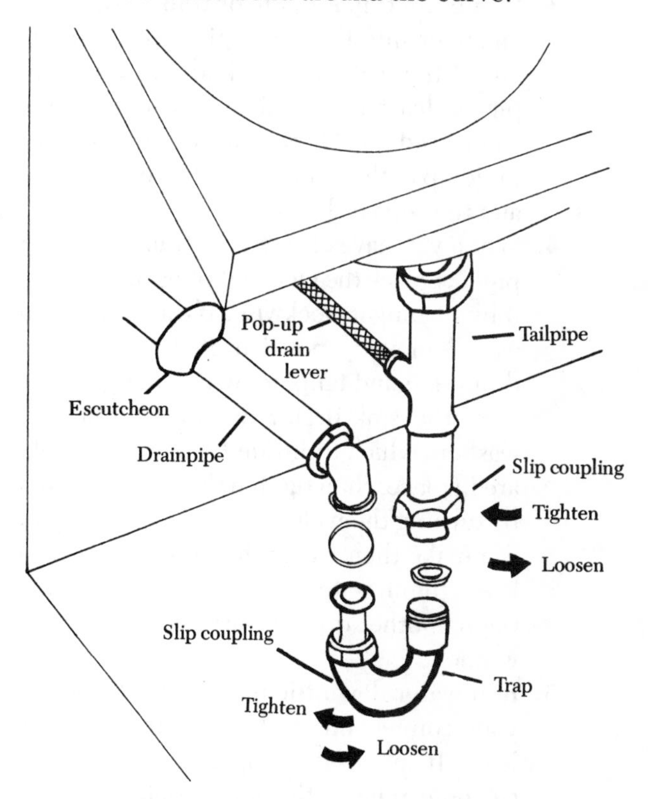

Anatomy of a trap. To remove it, simply loosen the couplings at either end of the curve.

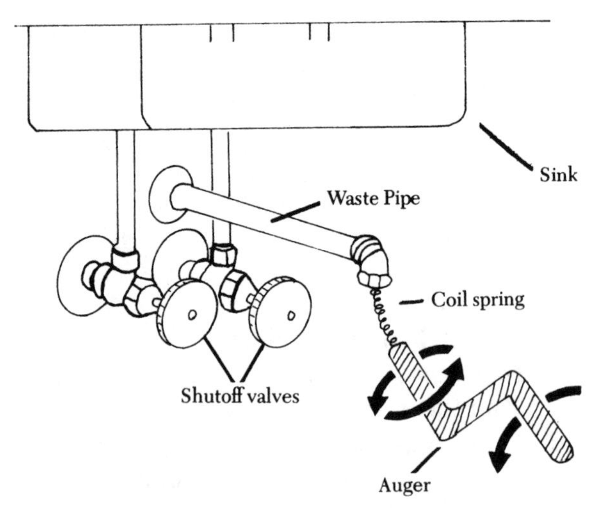

While the trap is removed, push an auger into the waste pipe to clear it of any obstructions.

3. While the trap is open (or removed), push an auger into the waste pipe that extends out of the wall or floor, and clean out the pipe at least as far as the soil stack that it is connected to. Often, an obstruction will build up at the joint between the drain line and the soil stack.
4. When you have cleaned the trap and waste pipe, replace the cleanout plug and tighten it by turning it clockwise. If the entire trap was removed, position it between the drain line and tailpipe, which hangs down from the sink drain. Be certain that the washers which fit inside the coupling nuts are in place, then tighten the coupling nuts by turning them clockwise as tightly as you can make them. That should be enough to keep them watertight. If it is not, give them another quarter turn with your pipe wrench.
5. Run water down the drain. Check each of your connections to be sure they do not leak. If any water appears, tighten the cleanout plug or the appropriate coupling nut with a pipe wrench until the leak stops.

Unclogging Bathtubs

When a bathtub drain becomes clogged, first try a plunger. Stuff the tub overflow port with damp rags, fill the tub with an inch or two of hot water, and then plunge the drain until it unclogs or it becomes apparent that your plunger will not work.

Your next approach is to clean the tub trap, which may be a standard S- or P-shaped trap under the drain—particularly if the tub is on the first floor of your house and access to the trap is from the basement. If you cannot locate a standard trap under or behind the tub, look for a drum trap. The drum trap looks like a tin can; it is made of steel or cast iron and has a screw-on cap which, if it has not frozen shut, can be undone with a pipe wrench. The drainpipe from the tub enters the top portion of the canister and continues out of the trap at its bottom. It may take years for the canister to fill with sludge and

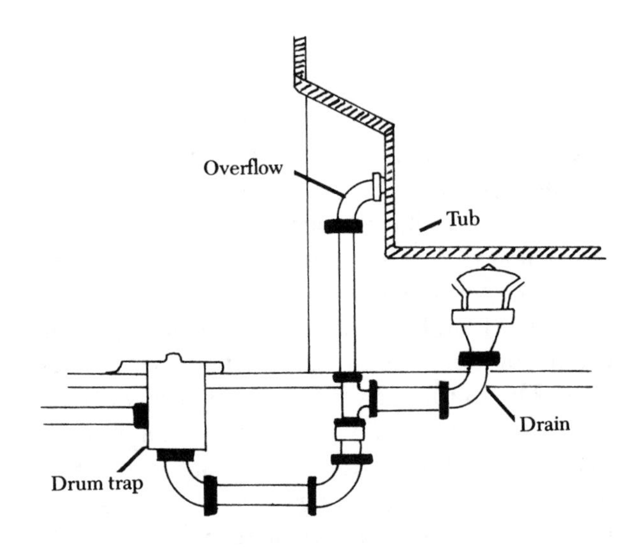

Drum traps are often found under hidden trap doors in the floor.

require a cleaning, but when this happens your only recourse is to reach into it and scoop out whatever is blocking the passage of water.

Drum traps are found in many older plumbing systems, and are no longer considered sanitary. Whenever possible, they should be replaced by a standard P or S trap.

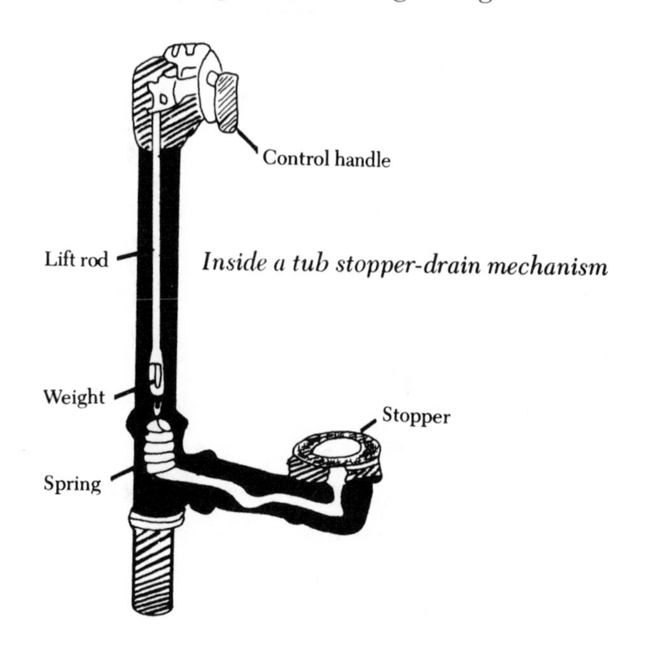

Inside a tub stopper-drain mechanism

STOPPER MECHANISMS

Although the trap under a bathtub is often the ultimate cause of stoppage, you may discover the problem involves the drain stopper. Stopper mechanisms usually have a spring that puts pressure on the rod assembly, which raises and lowers the drain stopper. Debris can collect around the spring and block the tub drain. If the stopper is a weighted type, dirt can accumulate on the weight and in the hole where it resides; this will prevent the stopper from completely closing and allow water to trickle down the drain. The procedure for disassembling and cleaning a tub stopper mechanism is:

1. Unscrew the plate around the base of the stopper lever.
2. Pull the lever mechanism out of its hole.
3. Clean all of the parts in the stopper mechanism thoroughly, paying particular attention to the bottom of the stopper.
4. Run hot water down the drain for several minutes in order to wash out any dirt that may have accumulated around the stopper seat.
5. Inspect all of the parts of the stopper mechanism for wear or breaks. If necessary, replace the entire mechanism.
6. Insert the lever mechanism in its hole and tighten the base plate in position.

REPAIRING FAUCETS

Problem *Possible Causes*	*Remedy*	*See Page*
	1. SPOUT DRIPS	
Faucet open	Close faucet	—
Handle defective	Check handle knurling, replace handle.	25
Stem defective	Check kurling, replace stem.	26
Stem washer defective	Replace washer	26
Seat worn	Replace or redress seat	27
2. SPOUT DRIPS, VIBRATES		
Faulty stem	Check stem, replace	25
Loose seat washer	Tighten stem washer	26
Defective assembly	Replace faucet	33
3. HANDLE DOES NOT CLOSE FAUCET		
Loose handle	Check knurling, replace handle	25
Stem defective	Check knurling, replace stem	26
4. WATER DRIPS FROM HANDLE		
Defective packing	Replace packing or washer	26
5. LEAK AT SWIVEL JOINT (MIXER FAUCET)		
Loose spout bonnet	Tighten bonnet	30
Defective spout packing	Replace packing	30
6. FAUCET LEAKS AT BASE (MIXER FAUCET)		
Defective mixing chamber	Replace faucet	33

REPAIRING VALVES

1. LEAK AT BONNET		
Loose bonnet nut	Tighten nut	35
Defective packing or worn packing washer	Replace packing or washer	35
2. VALVE WILL NOT CLOSE PROPERLY		
Worn seat washer	Replace washer	35
Defective stem	Replace valve	36

CHAPTER 2

Repairing Faucets

A FIXTURE, in plumbing parlance, is a receptacle for water. It can be a sink in your kitchen or bathroom, a basin in the laundry room, the bathtub, shower, or toilet. Every fixture except the toilet is serviced by two water supply lines, one for hot water and one for cold, and is also connected to the house drainage system so that water can be drained away from the fixture if necessary.

How much water and at what temperature it enters each fixture are controlled by the faucets, which constitute most of the moving parts in any plumbing system. But moving parts are susceptible to wear, which in turn causes leakage; it is this leakage which accounts for the majority of plumbing system repairs in any house. Most of the work you must do to maintain your plumbing system, then, is centered at the fixtures.

There are only a few repairs that can be made to any faucet. To make any of these repairs, you will need:

1 adjustable-end wrench
1 standard-blade screwdriver
1 phillips-head screwdriver
1 10" pipe wrench
assorted seat washers
friction tape

Faucets and Their Components

There are several different kinds of faucets, not only on the market but probably in your home. Some of them are washerless; some are single-lever diverter types or single-knob mixer faucets. Most are old-fashioned, standard seat-washer units. And each faucet has valves, which

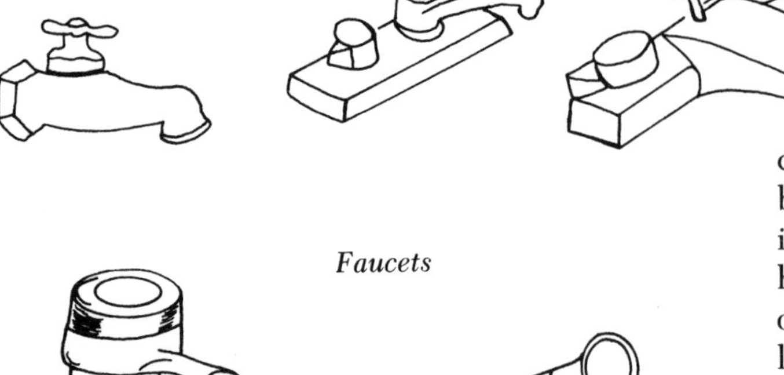

Faucets

function in precisely the same manner as the faucet itself except that they do not have a spout; water merely passes or does not pass through their bodies.

No matter what a faucet looks like, there are certain components that are sure to be found in its construction. The handle is always attached to a stem which extends into the faucet body. A screw is always placed in the center of the handle, but it may be hidden under a decorative cap which has the letter **H** (for hot) or **C** (for cold) on it. If there is a cap, it is held in place by metal

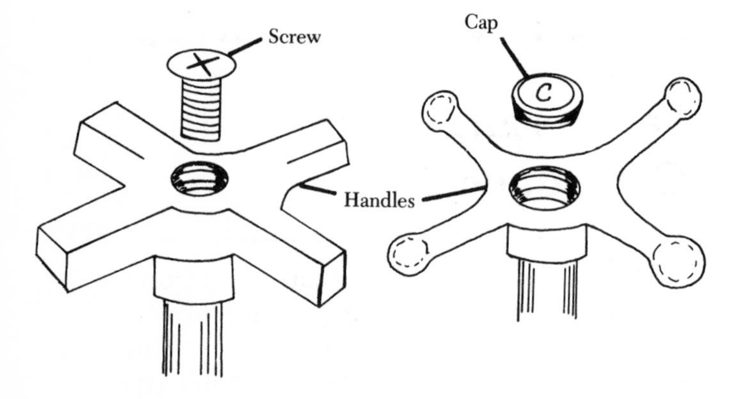

The screw fastening a faucet handle may be hidden under a cap that must be pried up.

clips and can be pryed off the handle with the blade of a screwdriver or knife. The screw fastening the handle can then be undone, but the handle will not necessarily lift off easily. The hub of the handle is knurled—that is, cut into interlocking vertical grooves so that even if the screw is lost, the handle may still be fastened to its stem to permit the flow of water through the faucet. To remove the handle, you may have to pry it upward, and it may resist you. But be assured that nothing is holding it in place other than the interlocking knurling.

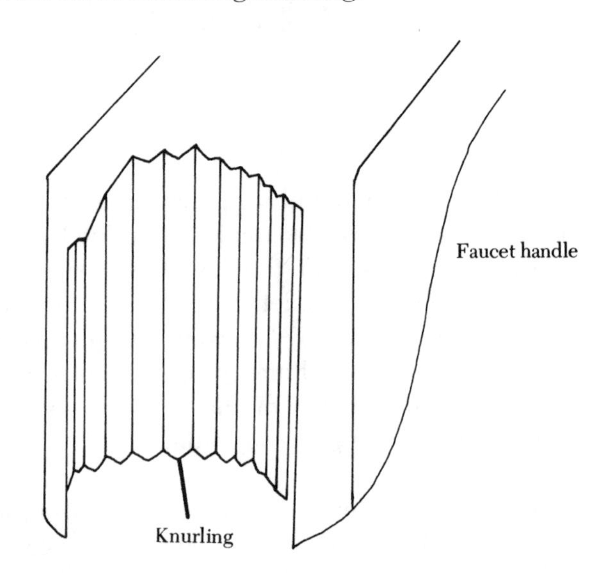

Knurling in a handle should be sharp and pointed, not worn or rounded.

A faucet stem may be threaded into the body that houses it, or it may simply fit very snugly. When disassembling the stem, always try unscrewing it first. If that seems to get you nowhere, pull it straight upward. You may find it is difficult to get a grip on it with your pliers or a wrench without damaging the knurling on the stem. But you can put the handle back on the stem, lock it in place with its screw, and then use the handle to pull the stem out of its body.

In order to prevent water from bubbling up the stem and seeping out under the handle, faucets must have a packing nut which covers the top of the faucet body and surrounds the stem. The nut ensures a watertight connection between the two parts by holding either a washer or a putty-like packing in place at the joint. Either may become worn and require replacement.

Most faucets have a seat washer attached to the bottom of the stem with a small brass screw. When water drips from the faucet spout, the culprit is a worn seat washer which must be replaced. Replacement requires that you dismantle the faucet handle, withdraw the stem, and replace the washer at its base with a new washer.

How to Disassemble a Faucet

1. Shut off the water supply valves nearest the fixture. Even if you are working on just one of the faucets, turn off both valves so there will be no chance of water "crossing over" to interfere with your work.
2. Open both faucets to allow all water in the pipes to drain off.
3. With the blade of a screwdriver or knife, pry off the cap covering the handle screw.
4. Remove the screw from the handle.
5. Pull the handle upward off its stem. If necessary, tap the underside of the handle with the base of your screwdriver. If you cannot get at the underside of the handle to tap it, you may be able to use the blade of your screwdriver to pry it upward. When you are prying or tapping the faucet handle, work around the base of the handle so that you can pull it evenly up the stem.

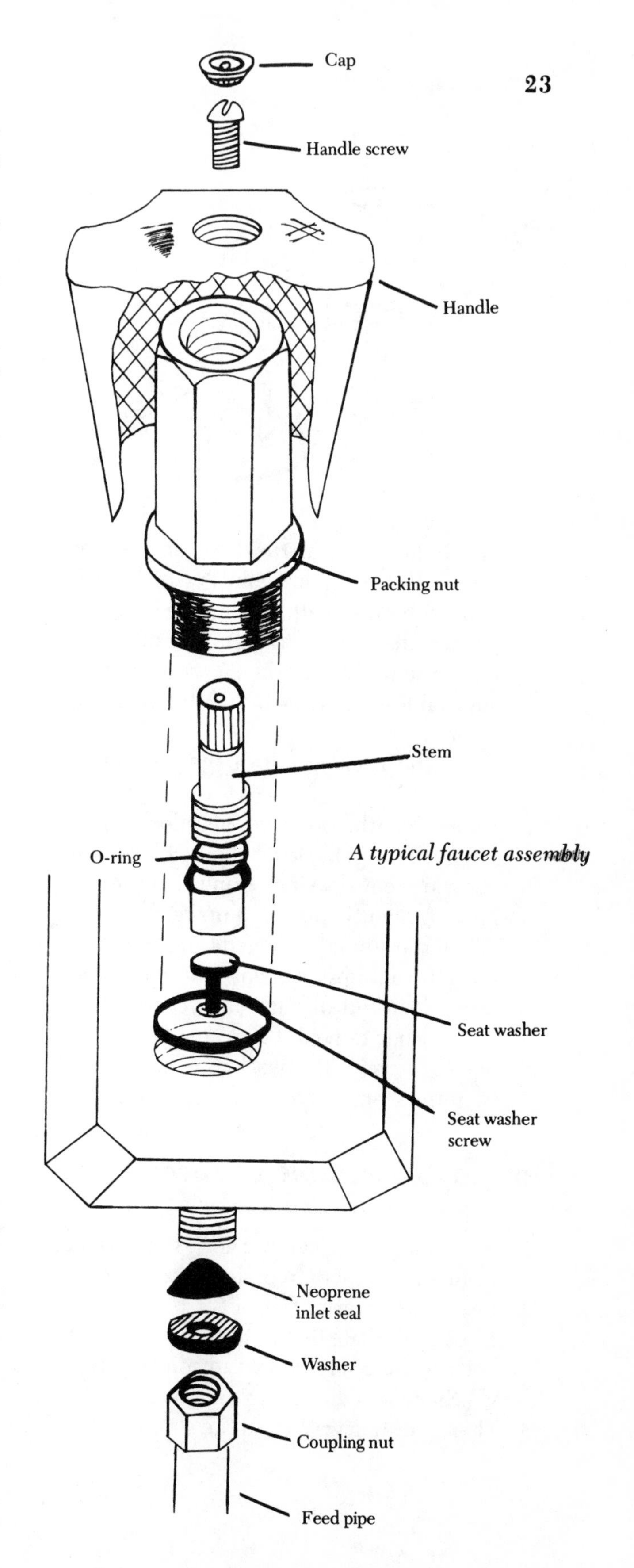

A typical faucet assembly

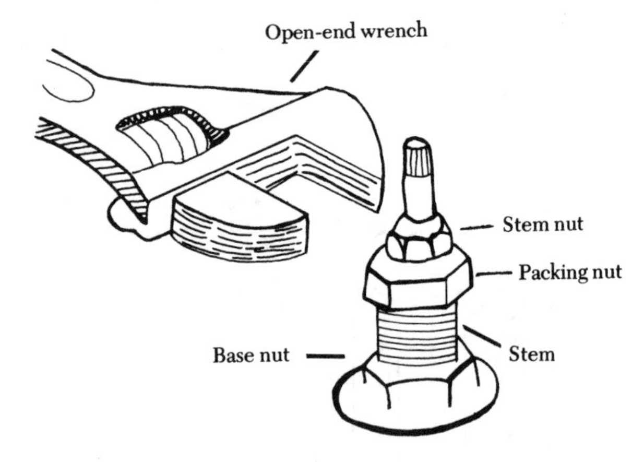

6. The bonnet nut at the base of the stem usually has a decorative finish which can be marred by the teeth of a pipe wrench. To protect the finish, wrap a piece of friction tape around the nut, then loosen it with an adjustable-end wrench and lift it off the stem.
7. Undo the packing nut at the base of the stem.
8. Hand-turn the stem counterclockwise to remove it from the faucet body. If unscrewing the stem does not bring it out of the body, try gently pulling it upward.
9. When the stem is removed from the faucet body, turn it upside down. There is a brass screw inserted through the center of the seat washer to hold it in place. By removing the screw, the washer will be freed from the stem.

How to Reassemble a Faucet

1. Be certain the seat washer is firmly attached to the bottom of the faucet stem.
2. Insert the stem in the faucet body. If it is threaded, rotate the stem clockwise into its body. If there are no threads on the stem, gently push it down into the body.
3. Place the stem washer on the stem, seating it squarely on the top of the faucet body. If there is packing, wrap it snugly around the stem. Tighten the packing nut over the washer or packing.
4. Put the bonnet nut over the washer or packing and hand-tighten it until it is snug, then tighten it with a wrench. Be careful not to overtighten the nut or you will strip the threads. If you find any leakage after the faucet is assembled, you can always tighten the nut more to stop the flow of water.
5. Put the handle on the stem and secure it in place with the handle screw.
6. Insert the decorative cap in its recess.
7. Open the shutoff valves.
8. Inspect the faucet while turning it on and off for any sign of leakage, either from the spout or at the bonnet nut. If there is no leak, remove the friction tape from the bonnet. Otherwise, disassemble the faucet far enough to tighten the packing nut.

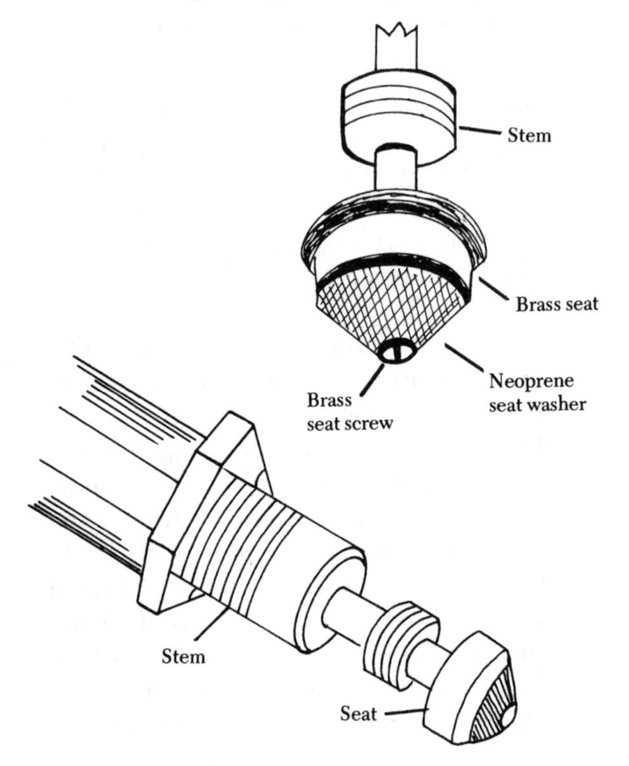

Faucet Repairs

Loose or Damaged Handle

1. Remove the handle from its stem (see page 23, steps 1—5).
2. Inspect the handle closely. If the knurling is rounded—that is, worn—or if any part of the handle is broken, it is best to replace the handle with a similar make. You can buy replacement handles at most hardware and all plumbing supply companies.

Interim Repair

In the event it is not immediately possible to replace the handle, wrap friction tape around the stem and push the handle down over the tape. If the handle is still loose, remove it and wrap more tape around the stem until the handle remains in its proper position. Eventually, the tape will wear down and the handle will become loose again, so the handle should be replaced as soon as possible.

Leak at Bonnet

1. Remove the handle and bonnet nut (see page 23, steps 1—6).
2. If a packing washer is used in the faucet, inspect it for wear. A faulty washer should be replaced with an identical one. If the faucet has packing, unwind it from around the stem and discard it. Replacement packing is sold at hardware and plumbing supply stores. Clean the stem of any residue from the old packing and then wrap the replacement packing around the stem.
3. Thread the packing nut over the packing, tightening it enough to compact the packing. Do not overtighten or the stem will be very difficult to turn.

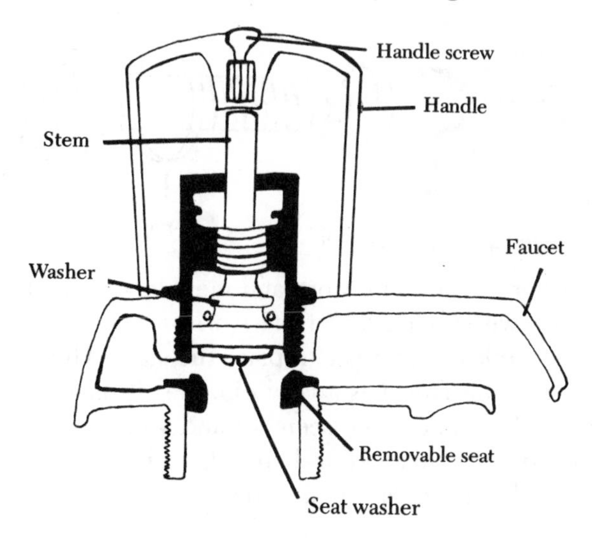

The inside of a faucet

4. Thread the bonnet cap over the packing nut. Replace the faucet handle and test the faucet to be sure there is no leakage. If any water appears around the bonnet cap, remove it and tighten the packing nut slightly.

Faulty Stems, Washers, and Faucet Seats

The faucet stem resides inside the body of the faucet and is rotated by the handle attached to its top to open and close the tap. The bottom of the stem contains a washer that fits into the faucet seat to prevent water from passing through the faucet when the stem is closed to its *off* position. When the faucet is open, the seat is lifted out of its place by rotating the handle and therefore the stem.

The knurling on the top of the stem can become so worn that the faucet handle cannot remain locked to the stem. The thread around the center of the stem can also become worn, making it difficult or impossible to rotate the stem in its body. The stem itself can become bent, and its

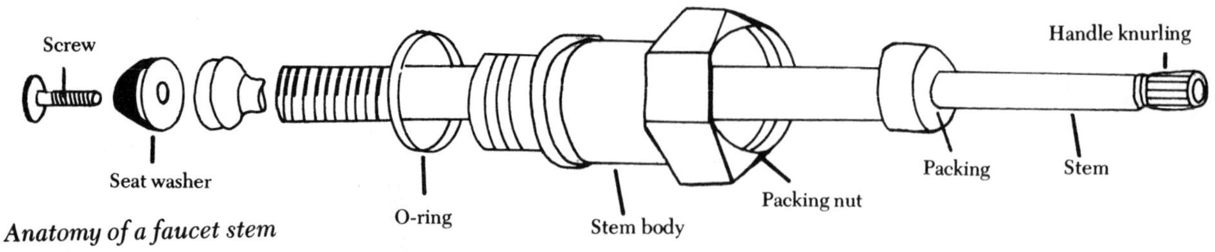

Anatomy of a faucet stem

washer can become worn and ineffective. In all cases, with the exception of a worn stem washer, the simplest repair is to replace the stem. However, if the faucet is a very old one, it may be difficult to find a replacement and you may be forced to install an entirely new faucet (see page 33 for faucet replacement procedure).

DEFECTIVE STEMS

It is wisest to remove the stem from its body and bring it to a plumbing supply outlet so that you can compare it to the replacement you are buying. It also helps if you know the manufacturer of the unit, although faucets are usually not marked. Sometimes you can find the sink manufacturer imprinted on the underside of the unit, but that is only a clue and does not guarantee that the faucets were made by the same outfit.

1. Disassemble the handle and bonnet, and remove the stem from the faucet body (see page 23, steps 1—6).
2. Examine the stem closely. If the knurling is worn, or the unit is bent, or the threads around its body are imperfect, replace the stem with an identical unit.
3. Insert the replacement stem in the faucet body and hand-turn it until it is secure.
4. Place the packing washer or packing around the top of the stem.
5. Tighten the packing nut over the packing or washer.
6. Tighten the bonnet over the packing nut.
7. Attach the handle and close the faucet.
8. Turn on the water supply shutoff valve.
9. Open and close the faucet to test for leakage.

DEFECTIVE SEAT/STEM WASHERS

If a faucet drips water when it is closed, the problem is, 99% of the time, a faulty seat washer (stem washer). Otherwise it is a defective stem. The seat washer is attached to the bottom of the stem by a brass screw driven through its center, and the washer itself is often partially contained in a small rimmed disk.

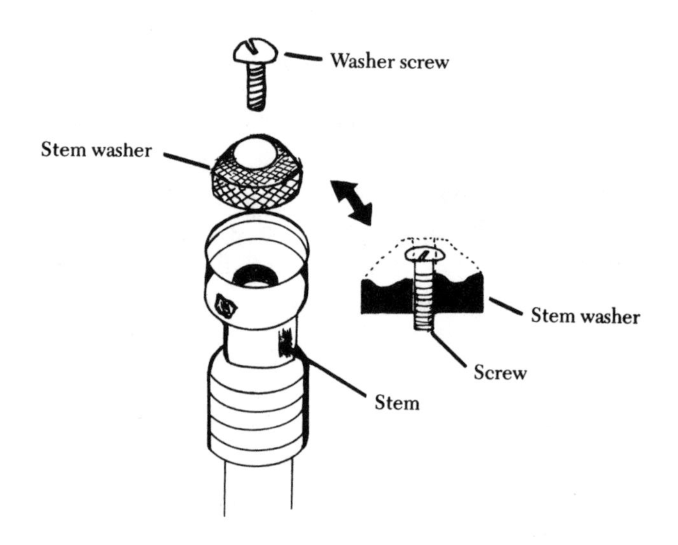

Seat washers are D-shaped, not flat.

Replacement seat washers can be purchased at almost any hardware store. They come in a plastic box or bag containing an assortment of sizes and there are usually three or four brass screws as well. Seat washers are unusual; they are not flat disks with holes in the middle of them, but are D-shaped. When you are replacing a seat washer, place the flat side against the metal disk in the faucet stem; the washer should be large enough to fit snugly inside the rim of the disk.

1. Remove the faucet stem (see page 23, steps 1—6).
2. Inspect the seat washer at the bottom of the stem. If it is chewed, or severely compressed, or looks very thin, or even if you notice nothing wrong but the faucet drips when it is shut off, the washer should be replaced. Undo the brass screw holding the washer in place and discard the washer.
3. Select a washer that fits snugly within the rim of the washer base plate. Tighten the brass screw through the center of the washer. This will flatten the washer somewhat, pushing its sides outward to form a watertight seal with the faucet seat whenever the stem is closed to its *off* position.
4. Replace the stem (see page 24 Steps 1—8).

DEFECTIVE SEATS

A faucet seat is essentially a hole in the base of the faucet body that is opened or closed by the washer attached to the bottom of the stem. Some faucets have removable seats which are simply screwed into the faucet body and can be easily unscrewed and replaced. But most faucets do not have a removable seat, so you will have to either "dress" the seat if it becomes scarred or out of shape, or replace the entire faucet.

Using a Seat Dresser

The **seat dresser**, sometimes called a seat grinder, is an inexpensive specialty tool consisting of a screw and handle and a series of different-sized cutting heads capable of smoothing the sides of faucet seats. The dresser is not a tool you will need very often, if ever, so buy one only if you have a defective faucet seat.

1. Disassemble the faucet handle, bonnet, and stem. (See page 23, steps 1—9).
2. Inspect the seat closely. If it is scarred or rough, it can prevent the seat washer from forming a watertight seal, or destroy it altogether. If the seat is removable, unscrew it from the faucet body and replace it with an identical part from a plumbing supply outlet.
3. To redress a faulty seat, place a seat dresser in the seat so that the refacing cutter at the bottom of the tool is firmly in the seat hole.

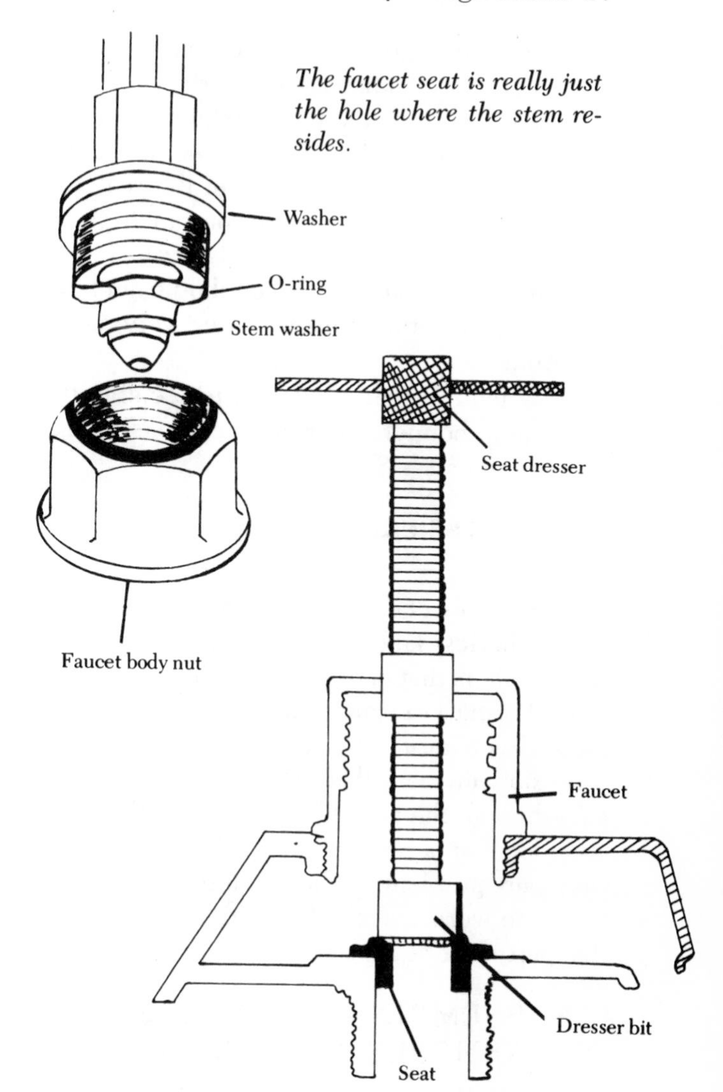

The faucet seat is really just the hole where the stem resides.

Seat dressers are used to smooth the inside of a faucet seat.

Hold the dresser vertically and steadily and rotate its handle so that the cutter can evenly scrape away the metal sides of the seat. Make only a turn or two at a time, then pull the dresser out of the seat and inspect the cavity. You do not want to shave off any more metal than is necessary to make the seat smooth; you should not have to rotate the dresser cutter very many times.

4. When the seat is smooth, reassemble the stem, bonnet, and faucet handle (see page 24 steps 1—8).

Washerless Faucets

Modern faucet designs include two versions of the washerless faucet, which, indeed, does not have washers that must be replaced from time to time. Washerless faucets operate by rotating a metal disk over a stationary disk, both of which have slots in them that align to allow water to pass through the faucet body. Because the parts are metal or a space-age plastic, they rarely cause any problems. When an internal part becomes so worn that the faucet develops a leak, that part must be replaced.

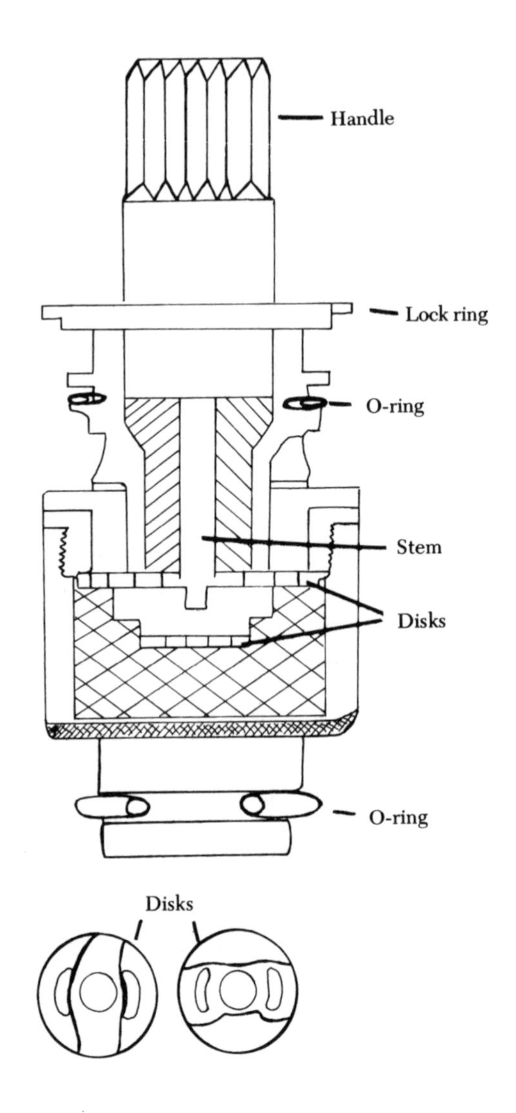

Components of a washerless faucet

DISASSEMBLY AND REASSEMBLY OF WASHERLESS FAUCETS

1. Close the water supply shutoff valve controlling the faucet.
2. Remove the faucet handle. You may have to take off a decorative cap in the center of the handle to locate the fastening screw, or it may be recessed elsewhere in the handle.
3. There is almost never any packing in washerless faucets. The nut under the handle is used to hold the internal parts together. Undo it with a wrench, and pull the stem out of the faucet body. Some of the washerless faucets have an O-ring seal around the middle of their stems, which can make the job of pulling the stem out of the faucet very difficult. The stem is not threaded, though. If you have difficulty withdrawing it, replace the handle and use that to pull the unit free.
4. Some disk-type faucet stems are held together by a pair of long screws inserted through the top of the stem and threaded into the body. To disassemble this type of faucet, undo the screws and pull the stem upward out of the body.

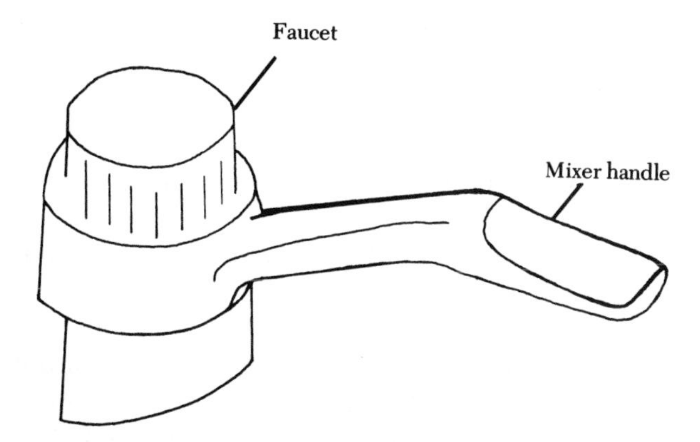

5. Although the moving parts of washerless faucets rarely wear out, chemicals in the water may deteriorate the rubber seals around the water ports in some models. Simply remove the remains of a defective seal and replace it with a similar unit, then reassemble the faucet.

Single-Handle Mixer Faucets

Single-handle faucets are designed to combine the hot and cold water that enters them through separate supply lines into the desired warmth before it leaves its spout. The combining is done by turning a mixing valve which may be controlled by a rotating handle or a lever. Most often, the combination faucets found in kitchen sinks have the lever, while bathtub, sink, and shower mixers feature a rotating handle.

There is nothing very different about combination faucets except that there is a diverter-valve assembly in the center of the unit. The diverter valve has a number of small parts that can usually be replaced if they become faulty, although it is most practical to simply replace the entire assembly. Fortunately, neither the diverter nor the handle assembly causes trouble very often.

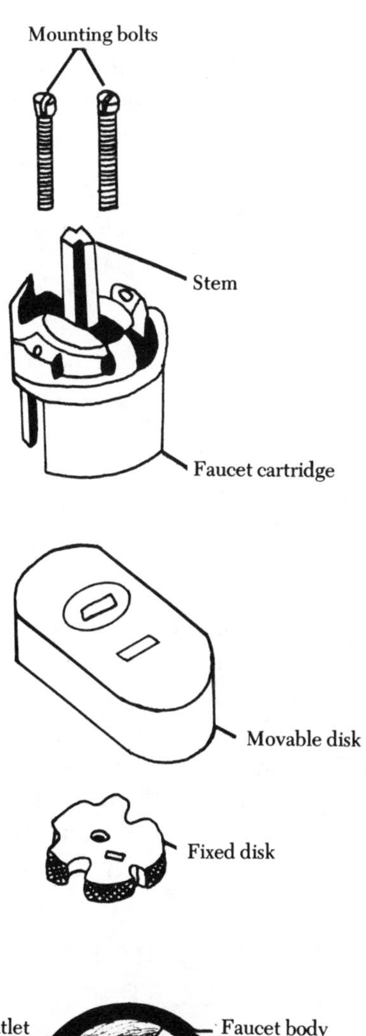

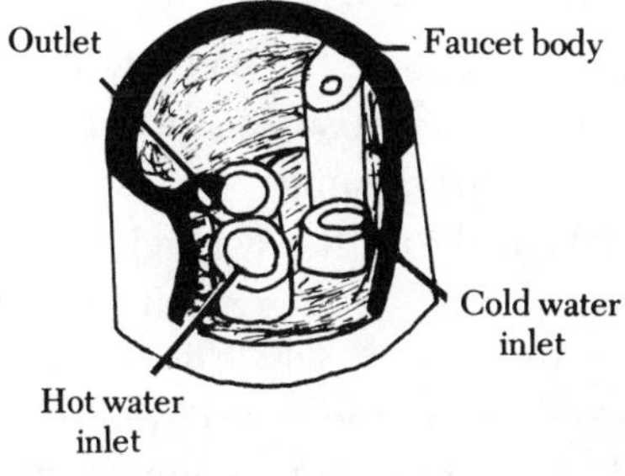

Components of a disk-type faucet stem.

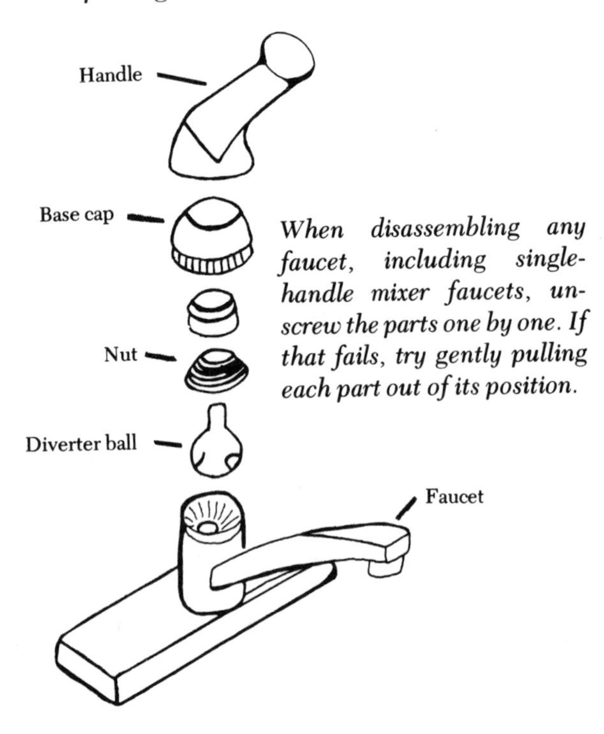

When disassembling any faucet, including single-handle mixer faucets, unscrew the parts one by one. If that fails, try gently pulling each part out of its position.

In order to repair any of the diverter faucets, you must have the manufacturer's repair sheet, which shows each of the parts and its proper position within the assembly. If you do not have that sheet and your mixer faucet requires extensive repairs, write to the manufacturer for a copy of the instructions. You might also try a plumbing supply outlet that sells the unit you own.

STOPPING LEAKS IN A MIXER FAUCET

Leaks can occur at the bonnet, at the base, and at the nozzle of a combination faucet, and are repaired in this manner:

1. If the leak is at the swivel joint at the base of the spout, wrap friction tape around the spout bonnet, which is merely a serrated ring around the base of the spout.
2. Tighten the spout bonnet with an open-end wrench by turning it clockwise until the leak stops.
3. If tightening the spout bonnet fails to stop the leak, the packing around the base of the spout or the spout washer may be defective. Close the water supply shutoff valves attached to both the hot and cold water supply lines.
4. Undo the spout bonnet by rotating it counterclockwise with an open-end wrench.
5. Lift the spout out of its socket and examine the packing inside the bonnet and the packing washer. The washer in this case is probably an O-ring, which is not flat but a round rubber or neoprene ring. If the packing is inordinately compressed, or the O-ring is damaged in any way, replace it.
6. Reseat the spout and tighten its bonnet.
7. Open the water supply shutoff valves and turn the faucet on and off several times to check for leaks. Tighten the spout bonnet until the leaks stop.
8. If the leak is coming from the base of the faucet (between the bottom of the faucet and the sink), the mixing chamber is defective. Chances are that the chamber cannot be repaired and the entire faucet must be replaced.

DISASSEMBLY AND REASSEMBLY OF BATHTUB MIXER FAUCETS

Bathtub mixer faucets have a metal or plastic cartridge inserted in the faucet body and held in place by a retainer clip. Whenever this type of faucet develops a leak, the only repair that can be made is to replace the cartridge.

1. Close both the hot and cold water supply shutoff valves.
2. Pry off the decorative cap in the center of the handle.
3. Remove the handle screw and pull the handle off the cartridge stem.
4. Pry the escutcheon away from the wall.
5. Pull the stop tube off the cartridge stem.
6. There is a tiny U-shaped clip in the end of

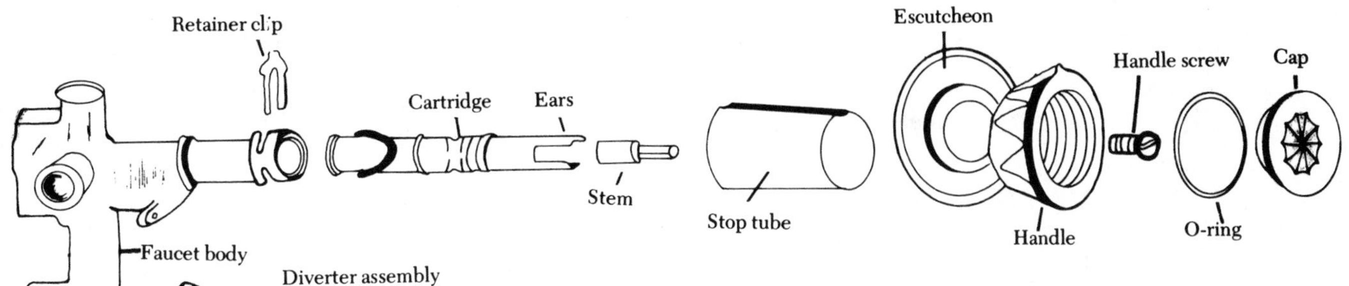

Anatomy of bathtub mixer faucet

the faucet body which is inserted through slots in the body. You may need needle-nosed pliers to grip the tab at the top of the clip so that it can be pulled upward out of its slots.

7. Grip the faucet stem and pull straight outward. The entire cartridge will come out of the faucet body. Take the cartridge with you to a plumbing supplier and purchase an identical replacement unit.
8. Insert the new cartridge in the faucet body. Rotate it until the tiny slots on each side of its ears are aligned with the retainer clip slots in the body.
9. Push the retainer clip down into its slots until it locks into place on either side of the body.
10. Slide the stop tube over the cartridge.
11. Place the escutcheon over the stop tube.
12. Attach the handle to the cartridge stem and tighten the retaining screw.
13. Press the decorative cap in the handle over the handle screw.
14. Turn on the water supply.

Sink Spray Hoses

Sink spray hoses are attached to a connector directly under the faucet spout. There is a diverter valve inside the base of the faucet which permits water to be rerouted from the spout to the hose when the sprayhead control is depressed. Insufficient water pressure may cause the hose to operate improperly, or the aerator screens in the hose can become clogged. You can purchase replacement hoses almost anywhere plumbing supplies are sold and easily replace a defective unit.

DISASSEMBLY AND REPAIR OF SINK SPRAY HOSES

1. Undo the hex nut at the end of the hose that connects it to the base of the spout.
2. Pull the length of hose entirely out of its guide hole in the sink.
3. Inspect the hose for kinks or cracks. If the hose is damaged, you may be able to repair it by wrapping friction tape around the damaged areas. But a more lasting repair is to replace the hose.
4. The nozzle can be removed by undoing the nut at the base of the unit and withdrawing it from the hose. The only time you need to take off the sprayhead is if it has ceased to function properly and must be replaced.
5. The aerator unthreads from the nozzle of the hose and should contain a small screen. If water does not come out of the hose properly, it usually means that the screen has become clogged. You can clean the screen and put it back in the nozzle and this will frequently solve the leak problem. If the screen is damaged in any way, or so clogged that it cannot be cleaned, take it to a hardware store and buy a similarly sized replacement.

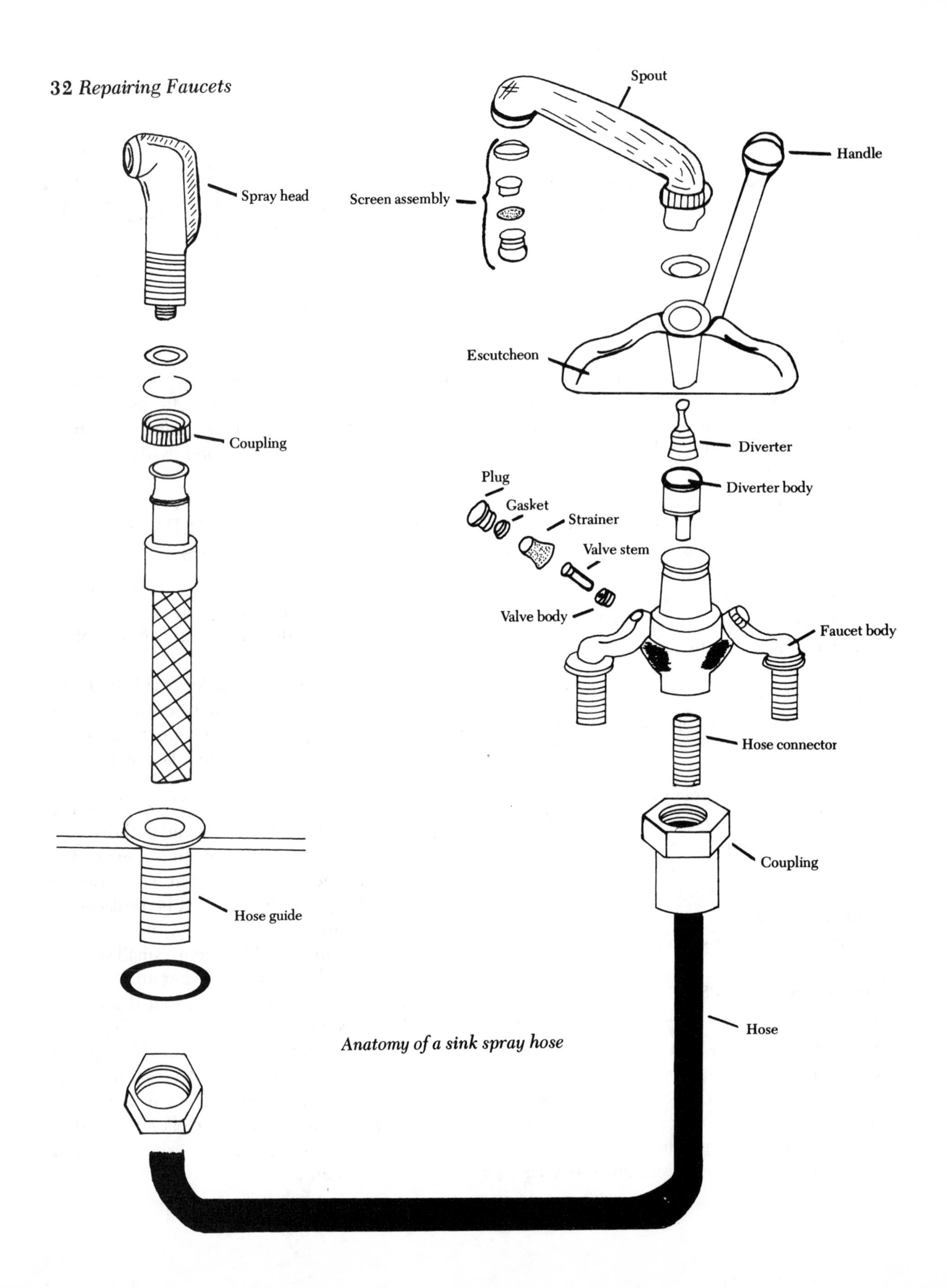

Anatomy of a sink spray hose

6. To reassemble the aerator, place the screen in the bottom of the aerator cap and insert the washer on top of it.
7. Screw the aerator back on the nozzle.

Replacing a Faucet

All faucets are basically installed and connected to the water supply system in the same manner, whether they are situated at the back of a sink or extend horizontally out of the wall behind the fixture. Replacing a faucet is not a particularly complicated task except for the fact that you have to work within the limited space of an inch or two between the back of the sink and the wall, or even behind the wall.

Because of the cramped quarters in which you must untighten and retighten nuts to make your water connections, you may need to purchase a basin wrench, which is designed to get you into tight spaces. You can assume that although changing a faucet is a relatively easy task, the limited space problem will make your working time somewhat longer than you might expect.

The tools you will need include an adjustable-end wrench, a 10″ pipe wrench, standard and phillips-head screwdrivers, plumber's putty, and plastic pipe-sealing tape.

1. Close both the hot and cold water shutoff valves and open the faucets to drain water from the pipes.
2. You will probably have to lie on your back under the sink, and you may need a flashlight to see behind the sink and locate the faucet tailpipes. The tailpipes extend down through the top of the sink and are threaded to accept couplings from the water supply line. Undo only the nut that holds the water line to the tailpipe. You do not need to tamper with any other connections.

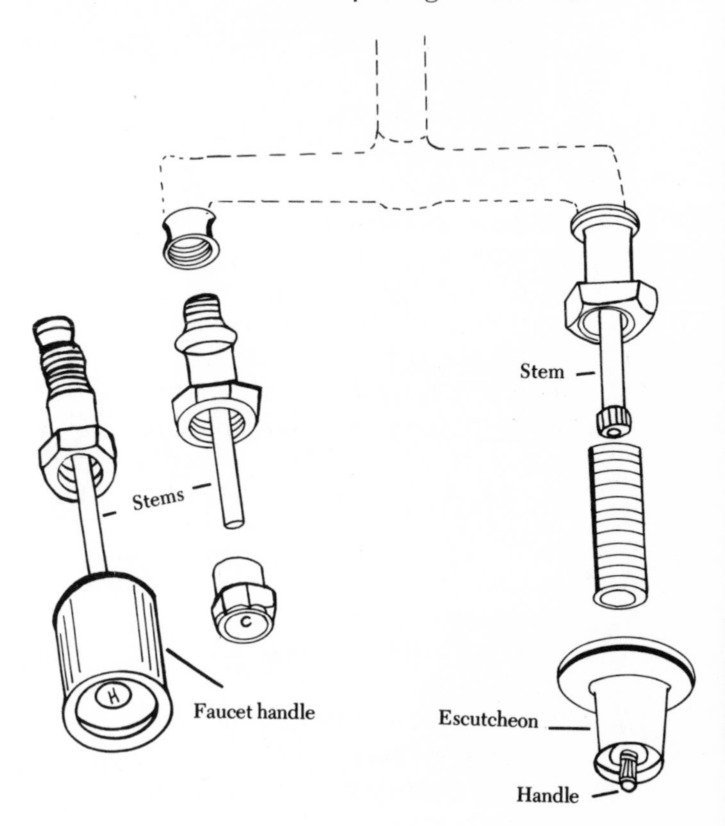

Don't be fooled by a faucet that comes straight out of the wall instead of up from a sink. All faucets are assembled in about the same way.

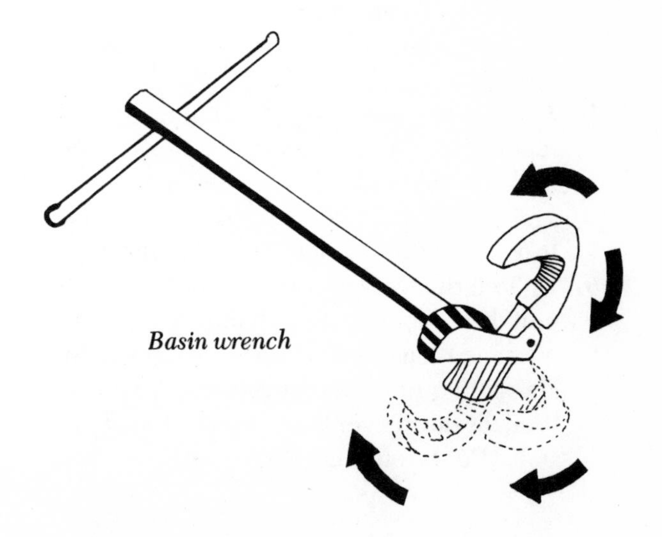

Basin wrench

3. Once the ⅜″ water supply line has been disconnected, it can be bent out of your way, but move it carefully to avoid breaking it. Either use a basin wrench to loosen the locknut, or place the end of a screwdriver against the locknut and tap it with a hammer. Be careful not to hit the locknut so hard that you damage the sink. When the lockwasher is free, remove it from the tailpipe by hand.
4. If both faucets are connected to the same spout, loosen the remaining faucet following steps 1—3. When all the pipe connections have been freed, you can lift the faucet off the sink.
5. Clean away all dirt, putty, rust, and any debris that has accumulated around the faucet holes in the sink. Dry the area thoroughly.
6. Remove the locknuts from the tailpipes on the new faucet and place them under the sink, where you can get at them when you need them.
7. Take a lump of plumber's putty and roll it between your palms until it is a long, even roll about ⅛″ to ¼″ thick. Wrap the roll around the edge of the faucet hole. If there are two holes, put a roll of putty around each hole. As an alternative, you can press the putty around the base of the faucet assembly.
8. Position the faucet with its tailpipe(s) extending down through the holes in the back of the sink. Press the unit down against the sink evenly until putty begins to ooze out from under the base of the unit.
9. Tighten the locknuts on the tailpipe. If the unit has two tailpipes, hand-tighten the nuts and then finish with a basin wrench by alternating nuts so that the faucet will be brought down evenly against the top of the sink. Be careful not to overtighten the locknuts or you may crack the sink or strip the threads on the tailpipes.

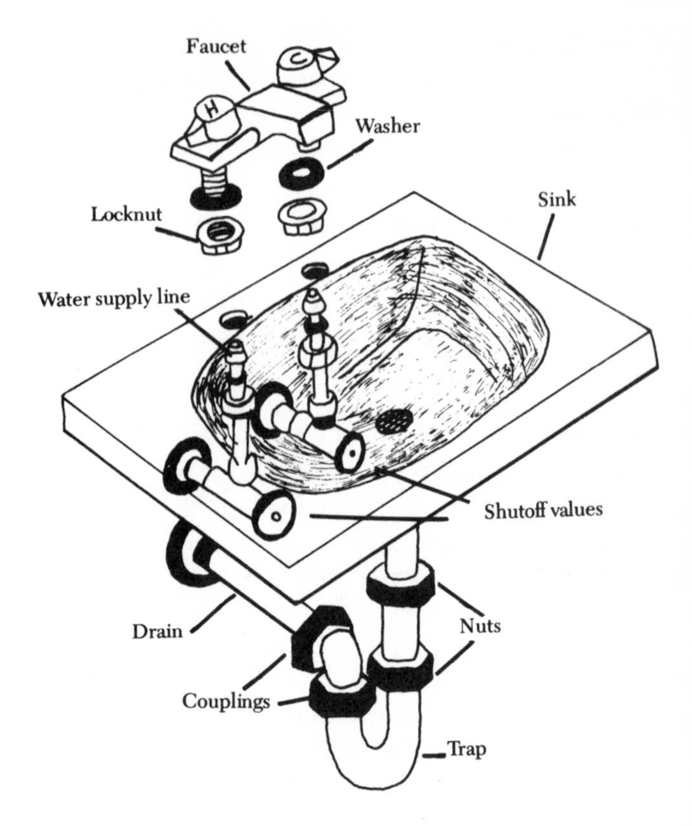

The major components of a faucet

10. Connect the water supply lines to the tailpipes. Wrap plastic pipe-sealing tape around the tailpipe in a clockwise direction and then tighten the supply line nuts on the tailpipe(s). Tighten them as much as you can without stripping the threads or splitting the nuts.
11. Wipe off any excess putty from around the base of the faucet.
12. If they are not already attached, put the handles on the faucet stems. The handle marked **H** goes on the left side, and the

handle marked **C** goes on the right. Lock the handles in place with their screws.

13. Turn on the water supply valves and open the faucets. Inspect all of your connections carefully to be sure they are not allowing any leaks. If there is a leak, tighten the appropriate nut until the leak stops.

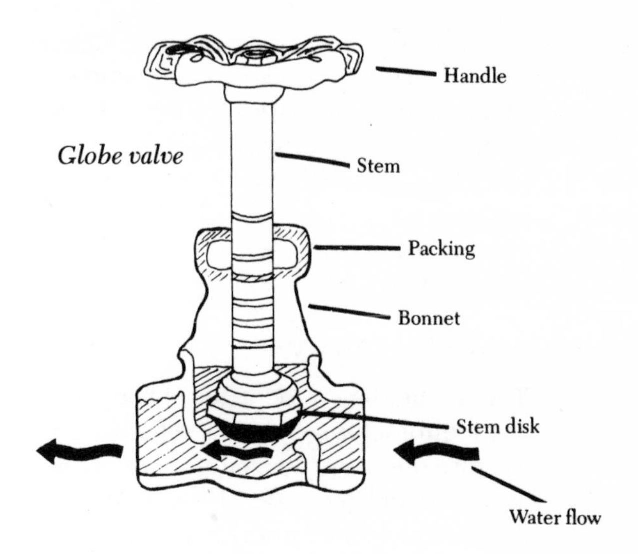

Repairing and Replacing Valves

All of the valves that are used in residential plumbing systems function in the same manner as any faucet. Since they have no spouts, turning their handles will drive the stem down into the base of the valve to close off the passage of water through the unit.

Valves have seat washers at the bottoms of their stems as well as packing washers under their packing nuts. Most often the valves are either the globe or the gate type. A **globe valve** is identifiable by the oval bulge under its base, and is designed to open so that the stem is entirely withdrawn from the base and water can pass freely through it. A **gate valve** has a series of half partitions in its base so that the water must twist and turn in order to get through it, thus slowing up the flow somewhat and reducing water pressure in the pipes. Gate valves are installed whenever it is desirable to lower the water pressure, such as where the main supply line enters your house or at some water-using appliances.

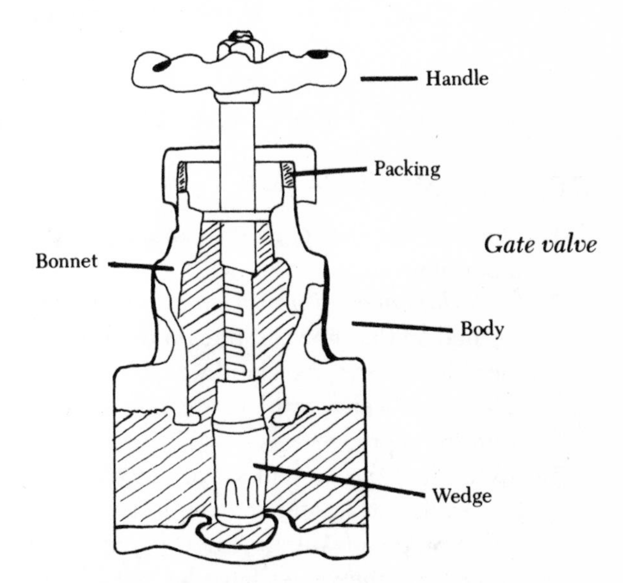

DISASSEMBLY AND REPAIR OF VALVES

1. No matter what type of valve is under repair, close off the water supply leading to the unit. This may mean you have to shut down the main service shutoff valve.
2. Open the faucets nearest to the valve and drain off the water remaining in the pipes.
3. Unscrew the handle and pull it off the valve stem.
4. Using a pipe wrench, undo the bonnet nut at the base of the stem.
5. Inspect the stem and inside of the bonnet nut. If the packing is compressed, it should

be replaced. Clean away all of the old packing and wrap new packing around the base of the stem, then replace the bonnet nut.

6. Unthread the stem by rotating it counterclockwise, and withdraw it from the valve body.
7. Inspect the seat washer. If it is faulty, replace it with a new one.

REASSEMBLING VALVES

1. Thread the stem into the valve body, turning it until it fits snugly in the body.
2. Be sure the bonnet washer or packing is in place around the base of the stem, then tighten the bonnet nut in place.
3. Place the handle on its stem and tighten the handle screw in place.
4. Turn on the water supply and inspect the valve to be sure there is no leakage.

REPLACING A VALVE

1. Close off the water supply and drain the pipes of water.
2. *For threaded pipe*—Using one pipe wrench on the pipe and one on the body of the valve, loosen the coupling nuts on each side of the valve until you can remove it from the pipes.
3. Insert the replacement valve between the pipe ends and tighten its coupling nuts.

   For Soldered Valves—If the pipes are copper and the valves have been soldered in place, you must melt the solder with a propane torch until the pipes can be pulled out of their sockets in the valve.

   New valves can be connected to the pipes in any number of ways: with flared couplings; soldering to the pipes; threaded couplings; adapters, and so on. The valve itself should be the same diameter as the pipe you are attaching it to; don't place a ½″ valve on ¾″ pipe, unless you intend to reduce the water flow once it is past the valve. If you are threading the valve in place, the "male" half of the connection (usually the end of the pipe) must be coated with a sealing tape or compound. If you are soldering the valve in place, the pipes have to be dry (follow the procedure for sweat-soldering given on page 86).
4. When the valve is in place, turn on the water supply and work the unit two or three times, checking carefully to be sure there is no leakage.

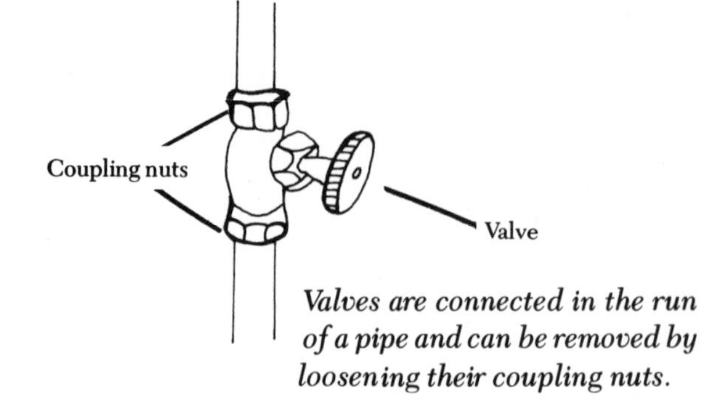

Valves are connected in the run of a pipe and can be removed by loosening their coupling nuts.

CHAPTER 3

Repairing Toilets

TOILETS REMAIN stoically in their corners year after year, apparently impervious to the world around them. In fact, about the only malfunction that develops in toilets involves the flush valve inside the tank, which can acquire a hiss, or fail to allow enough water into the tank, or prevent the toilet from flushing all of its waste.

Until about 1975, the typical flush valve was a complicated assembly of moving parts that open and close whenever the tank is emptied and needs to be refilled. Manufacturers continue to put the old-fashioned float ball flush valves in modern units, even though some of the new space-age replacements have proven themselves infinitely more reliable and durable.

The repairs that can be made inside a toilet tank include adjustment or replacing the flush handle, the inlet stopper, the flush valve, and the overflow tube. Some repairs are as simple as bending a metal rod; others demand the wholesale replacement of one unit or another. None of the work is difficult, but it can be frustrating and require some persistence—for example, an inlet stopper that does not close properly may require an hour or two of bending its different rods in all directions until the stopper seats properly.

You'll need the following tools for any or all of the toilet tank repairs:

1 open-end wrench
1 screwdriver, standard blade
1 chain or strap wrench (necessary only if you are working with the spud nut on the bottom of the tank)
1 10" pipe wrench
friction tape

Flush Valve Assemblies

The standard flush valve assembly includes a float ball which rises and falls with the water level in the toilet tank, causing the flush valve to open or close. Float balls were once made exclu-

REPAIRING TOILETS

Problem *Possible Causes*	*Remedy*	*See Page*
1. WATER RUNS CONTINUOUSLY		
Tank stopper does not seat properly	Bend guide rods or alter length of chain	46
Tank stopper defective	Replace stopper	47
Defective float ball	Service or replace float ball	40
Faulty inlet valve	Service or replace valve	41
Defective overflow tube	Repair, or replace tube	48
2. TANK DOES NOT FLUSH		
No water supply	Turn on water supply	—
Defective handle	Repair or replace handle	44
Defective handle or chain	Bend rods; repair or replace rods or chain	46
Faulty flush valve	Service or replace valve	41
3. TANK LEAKS		
Damaged tank	Replace tank	48
Defective flush valve	Repair or replace valve	41
Defective spud washer	Replace washer	41
Defective stopper	Replace stopper	47
4. TOILET LEAKS AT BASE		
Damaged bowl	Replace bowl	48
Defective grouting around bowl gase	Replace grouting	49
Loose bowl	Tighten bowl	49
Ferrule Gasket defective	Replace gasket	49
5. DAMAGED TOILET SEAT		
Broken seat	Replace seat and cover	54
Loose seat	Tighten seat nuts	54

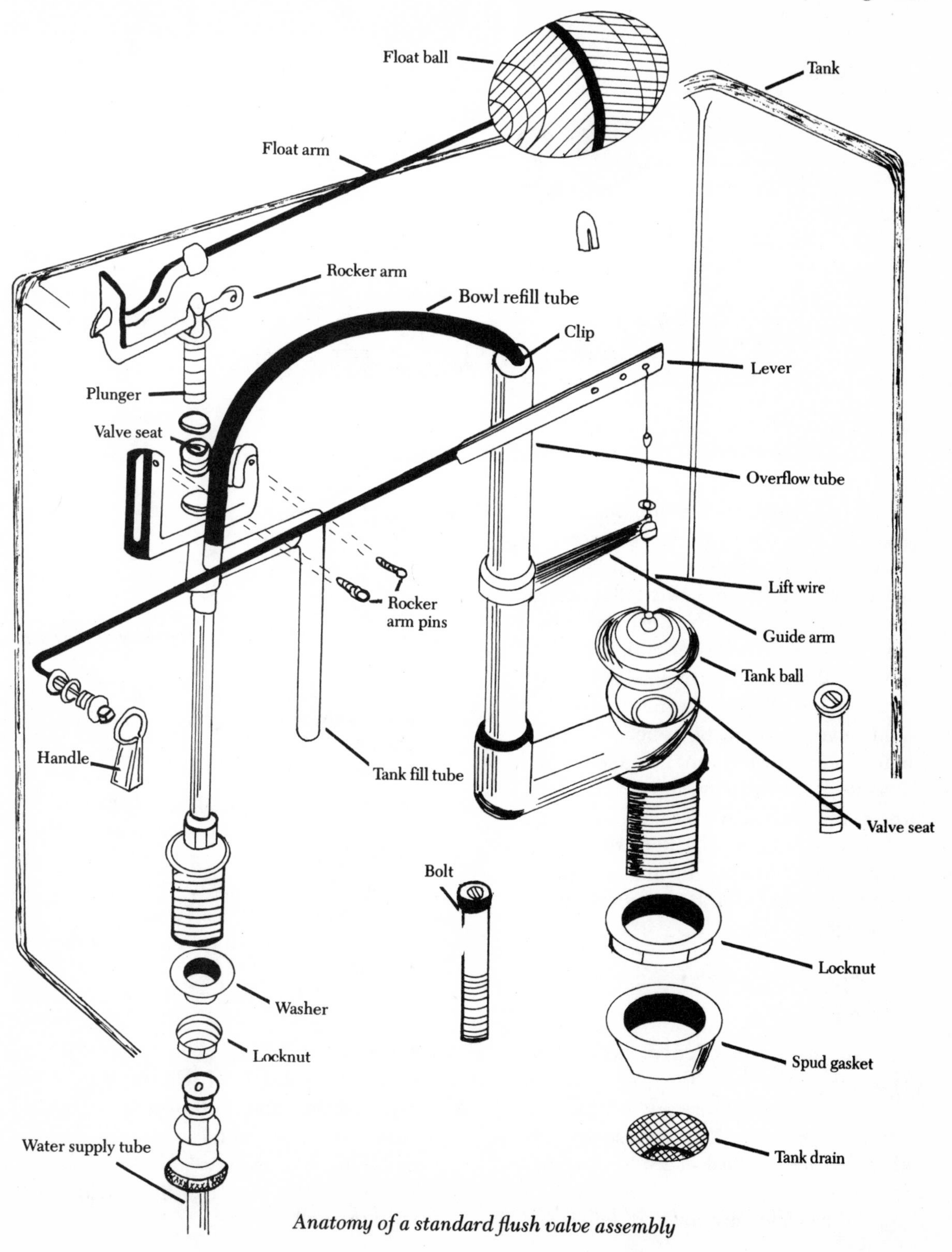

Anatomy of a standard flush valve assembly

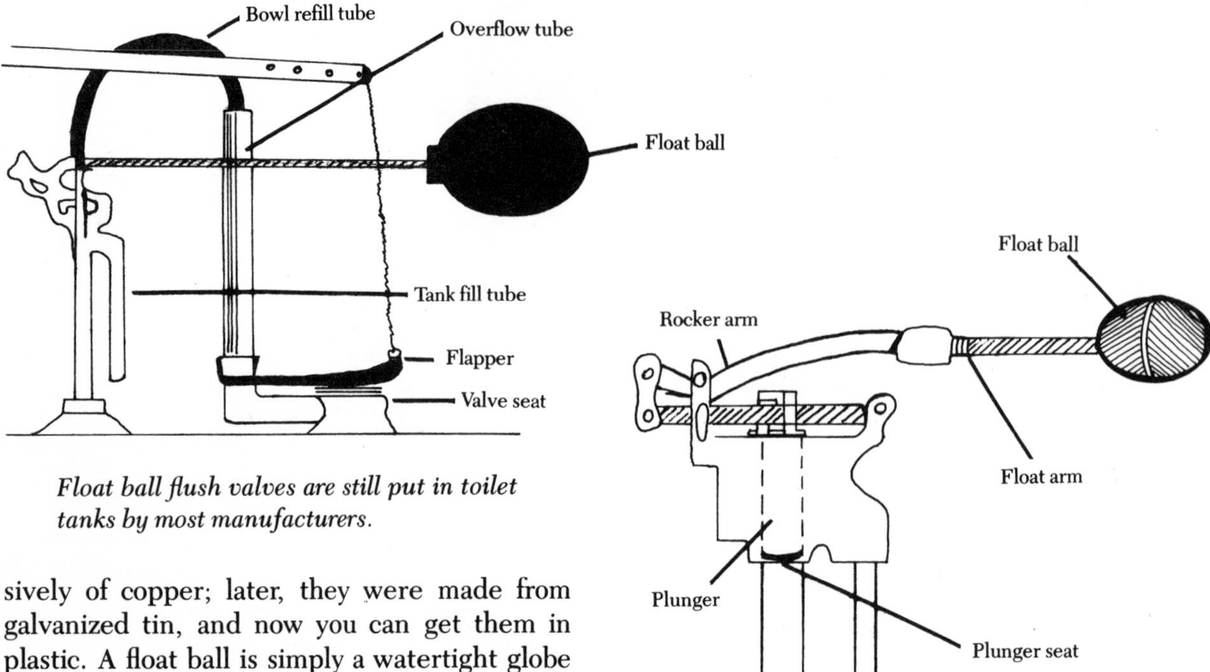

Float ball flush valves are still put in toilet tanks by most manufacturers.

The float ball arm threads into the valve rocker arm.

sively of copper; later, they were made from galvanized tin, and now you can get them in plastic. A float ball is simply a watertight globe that floats on top of the water in the toilet tank. It is connected to a rocker on the flush valve by a metal rod so that as the ball rises with the water, the plunger in the valve is driven slowly down until it seals off the fill inlet. In theory, the float rod should never have to be touched; but in practice, it is nearly always bent to make the ball close the value at the proper water level. Often, you can stop the flush valve from hissing (for a time, at least) by bending the float rod. But a more lasting remedy for a hissing flush valve is to replace the seat washer on the plunger or install an entire new flush valve.

TROUBLESHOOTING FLOAT BALLS

1. The float rod is threaded at both ends. Using pliers that grip the rod, you can rotate it counterclockwise and unthread it from the rocker arm. You can also grip the rod and rotate the float ball until it comes free of the rod.
2. Inspect the ball for any cracks or holes. If it is damaged, replace it. A float ball that fills with water will not open and close the flush valve properly.
3. Inspect the float rod to be sure its threading is not stripped. If the threads are in disrepair, replace the rod. If the rod has undergone a number of bendings and you cannot straighten it out, consider getting a new rod and starting all over again.
4. Thread the float ball on one end of the rod, rotating it until it is tight.
5. Screw the free end of the rod into the rocker arm.
6. Flush the toilet and allow it to refill. Observe the water level when the float ball shuts off the valve. If the level is too high, bend the center of the rod up so that the ball is aimed downward. If the level is too low, bend the center of the rod down to raise the ball.

Disassembling and Repairing Standard Flush Valve Assemblies

The following procedure can be used to dismantle and repair a standard flush valve, which you will find in 90% of the toilet tanks common today. The necessary replacement parts, even whole kits of them, can be purchased at most large hardware or plumbing supply stores.

1. Shut off the water supply valve located under the toilet.
2. Remove the toilet tank cover and put it aside.
3. Flush the toilet, holding the tank stopper up so that as much water as possible can drain out of the tank.
4. There is a rocker arm suspended over the valve seat by a rod; the rod is held to the assembly by a pin or set screw at each end. Remove the set screws and lift the arm assembly upward. You will pull the plunger out of its seat.
5. The plunger has a small washer screwed to its bottom. Inspect the washer for signs of wear and replace it if it is faulty. Undo the brass screw and pull the washer off the bottom of the plunger. Then insert the screw through the center of a replacement washer and tighten it in the plunger.
6. Use a thin wire to clean the small hole at the bottom of the valve seat and the inlet tube.
7. The fill tube hangs down from the underside of the valve seat. Remove it by carefully unscrewing it counterclockwise. The tube can be cleaned with a wire, then flushed with water.
8. Screw the fill tube back in place. The tube is relatively delicate; be careful not to bend or otherwise damage it.
9. If there is a washer around the top of the valve seat, inspect it for wear; if it is faulty, replace it.
10. Insert the plunger through the valve washer and into the valve seat. Position the rocker arm so that the set screws can be replaced.
11. Lock both set screws through the ends of the rocker rod. Work the rocker up and down to be certain it is properly installed.
12. Open the water supply and allow the tank to fill. The float ball should rise with the water and close the flush valve when the water reaches the proper level in the tank.
13. Flush the tank and allow it to refill again, just to be sure that all of the components are working properly.

Replacing a Flush Valve

Replacement flush valves are available at all plumbing suppliers and many hardware stores. You can buy either a plastic valve or, for considerably more money, one made of brass. Five dollars or so will buy one of the new fluid-level control valves or pressure-controlled valves that

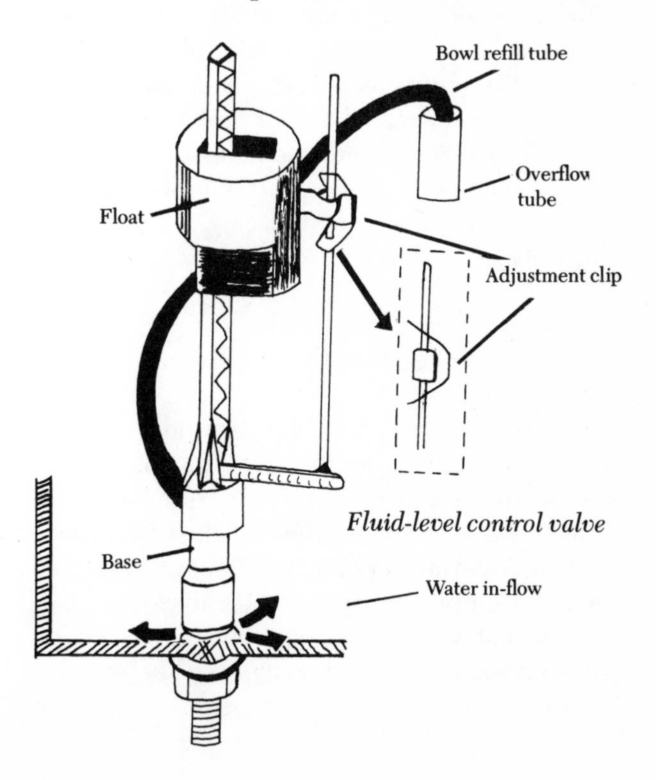

Fluid-level control valve

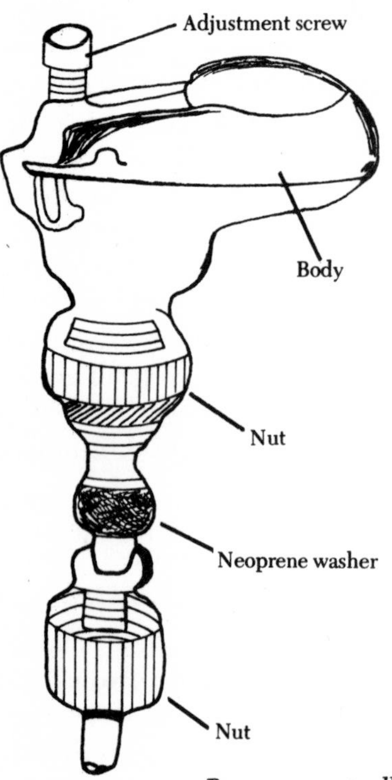

Pressure-controlled fill valve

have far fewer moving parts and therefore need almost no servicing. Either type of valve will replace the entire mess inside your tank, and eliminate most if not all of your flush valve problems. (The various components of flush valves are available both singly and in kit form, if you are intent on repairing or replacing your present assembly.)

Fluid-Level Control Valves

The FLC valves do not have a float ball or float rod, but are controlled by the pressure of water in the tank against a plastic float that rides up and down a plastic bar. The plastic is a tough, space-age material that will not corrode, rust, or yield to minerals in the water. Best of all, FLC units never hiss and are very precise about shutting off the water supply. Repairs to FLC valves occur so seldom that you cannot buy any replacement parts anywhere except directly from the manufacturer.

Pressure-Controlled Fill Valves

Even newer to the market are the tiny pressure-controlled fill valves, which are no more than 4½″ long and 5″ high. These are also made of durable aerospace plastic and function according to the amount of the water that weighs down on the top of the unit. The unit has one "repair" that can be made to it and this is fully described on the back of its package: there is a small rubber disk inside the cover which can be lifted out and washed. The cover is held to the body of the unit by set screws and also has a water-level adjustment screw protruding from it. The level of the water in the tank is controlled by rotating the screw.

No matter what type of flush valve you are working with, the installation procedure is the same:

1. Close the cold water shutoff valve positioned under the left side (as you face it) of the toilet.
2. Lift the cover off the tank. It is slightly fragile, so put it down carefully somewhere out of your way.
3. Flush the toilet and hold up the chain or rods that control the inlet stopper at the bottom of the tank so that as much water as possible will drain out of the tank.
4. Sponge out all of the water remaining in the bottom of the tank. If you don't bail out all of the water, it will come out of the tank through the valve port when you remove the flush valve assembly.
5. Undo the nut under the tank that connects the water supply line to the flush valve tailpipe. The feed line is often flexible copper tubing and you can bend it out of your way if you need to.
6. Unthread the locknut around the tailpipe. The nut is up against the underside of the tank and holds the flush valve vertically in the tank.
7. When the locknut is free of the tailpipe, you can lift the flush valve and its float ball

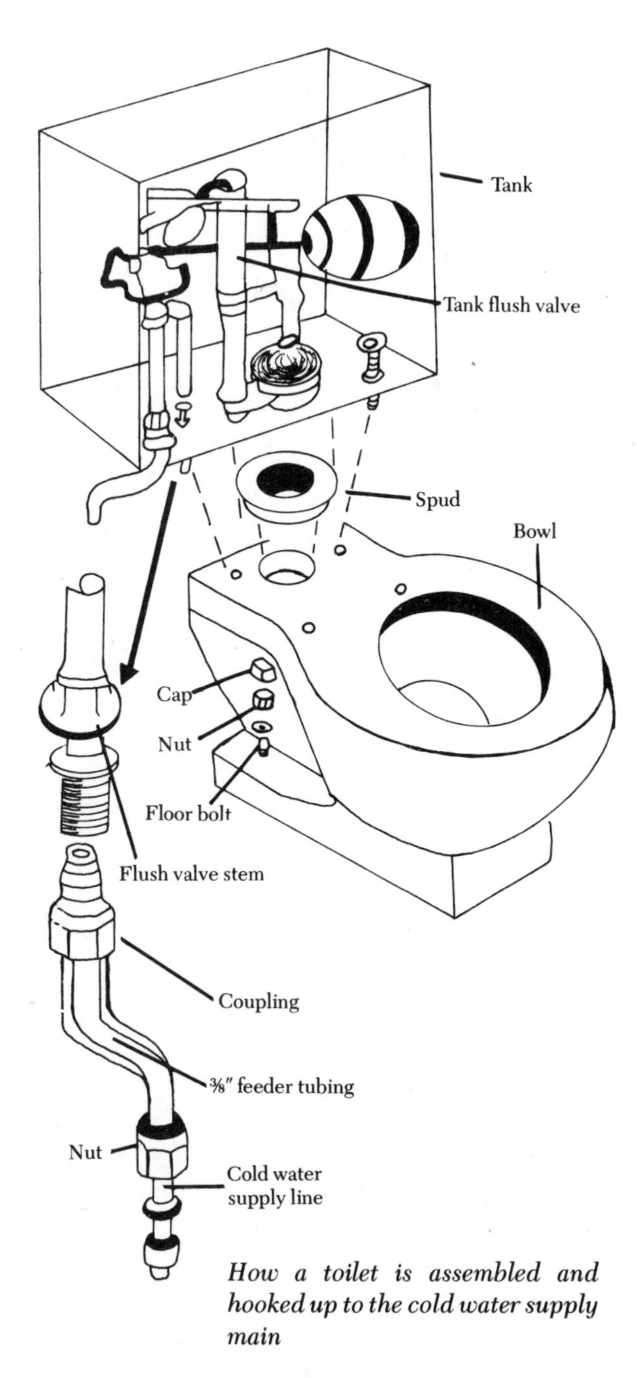

How a toilet is assembled and hooked up to the cold water supply main

out of the tank. There is a small rubber refill tube which leads from the valve to the top of the overfill tube, where it is held in place with a spring metal clip. Pull the clip off the overflow tube.

8. Assemble the neoprene or plastic washer on the tailpipe, pushing it up against the flange at the bottom of the unit.
9. Stand the unit up in its hole in the bottom of the tank.
10. Hand-tighten the locknut to the tailpipe. You should be able to get it tight enough to keep the flush valve upright, but you will have to hold the assembly in place with one hand and reach under the tank to tighten the nut with the other. When the valve is in its proper vertical position, tighten the locknut with your pipe wrench about half a turn. You do not want to risk stripping the threads on the tailpipe or cracking the tank; if there is a leak at the locknut when you test your connections, you can tighten it more.
11. Connect the water supply riser pipe to the tailpipe of the flush valve. Most replacement valves come with two or three different nuts and washers and include manufacturer's directions that tell you which ones to use with the water supply tube you happen to have (the tube may have a coupling nut and flared end, or it may be straight). You will be able to use whatever coupling arrangement is on the tube with the new valve unit. Place the appropriate washer on the end of the supply tube, wrap plastic pipe-sealing tape clockwise around the end of the tailpipe, and hand-tighten the supply tube nut to the tailpipe. Tighten the nut half a turn farther with your pipe wrench.
12. Attach the rubber or plastic refill tube to the nipple on the flush valve and clip it to the overfill pipe.
13. If you are installing a float ball flush valve, unscrew the ball and its float rod from the old unit and screw it on the new assembly. If you are putting in either of the new valves, you do not need the float ball or its rod.

14. Replace the tank top and turn on the water supply valve. Check your connections for any leakage. Water may appear at the nut between the tailpipe and supply line or the locknut under the tank. In either case, tighten the nut slowly, until the leakage stops.
15. Allow the tank to fill, then remove the tank top. The water level should be adjusted so that it is approximately ¾" below the top of the overflow pipe:

    Float ball valves—Adjust the water level by bending the float ball rod upward to raise the water level and downward to lower it.

    FLC valves—There is a clip on the float guide rod. Squeeze the tabs of the clip together to move it up or down on the rod until the float stops the water intake at the desired level.

    Pressure-controlled valves—These have a small knob on their top which is rotated clockwise to raise the water level and counterclockwise to lower it. The easiest way to adjust them is to let the water in the tank attain the desired level, then reach into the water and turn the knob until the unit shuts off.
16. When the water is at the proper level, replace the top of the tank.

Flush Handles

The flush handle resides on the front of the tank and is connected to a trip lever inside the tank. The trip lever raises or lowers the ball or stopper over the drain. It is important that the ball or stopper settles squarely into the drain; if it does not seal the drain, water will trickle out of the tank and into the bowl in a never-ending stream that is not only wasteful but annoyingly noisy.

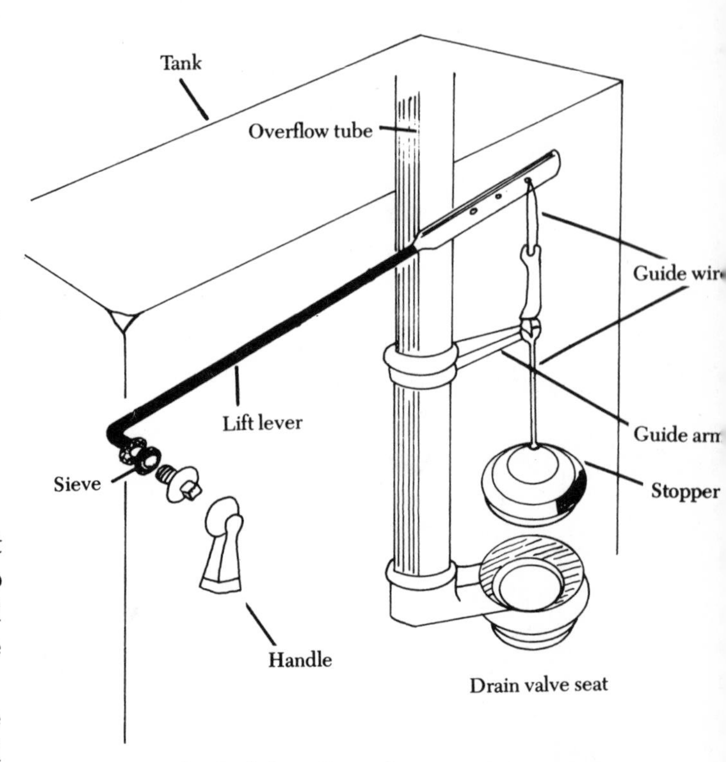

Anatomy of a flush handle and stopper-lift mechanism

Repairs to flush handles are as follows:

1. Shut off the water supply valve.
2. Remove the tank cover and flush the toilet.
3. Examine the flush handle and the trip lever for any sign of wear or damage. If either is worn or cracked, the unit should be replaced by an identical model. You can purchase replacement units at most hardware or plumbing supply stores.
4. To remove the handle and trip lever, first disengage the chain or lift rods from the end of the trip lever.
5. Using pliers, undo the nut which holds the trip lever to the back of the handle.
6. Remove the nut and washer that are around the tailpipe of the handle, on the inside of the tank.

7. Pull the flush handle out of its hole in the front of the tank.
8. Insert the new handle through the handle hole. There should be a washer between the face of the tank and the handle.
9. Place a washer on the handle tailpipe and hand-tighten the locknut against the inside of the tank. Finish tightening the nut with pliers, but only enough so that the handle is secure and still able to work freely.
10. Insert a washer over the threaded shank on the handle and place the end of the trip lever on the same shank. Sandwich the trip lever between another washer and tighten the lever nut on the shank to secure the assembly.
11. Connect the stopper-lift assembly to the free end of the trip lever.
12. Open the water supply valve and fill the tank.
13. Depress the handle to empty the tank. Watch the handle and trip lever to be sure they are functioning properly; if you discover that the handle sticks, loosen the nut holding the trip lever and handle together.

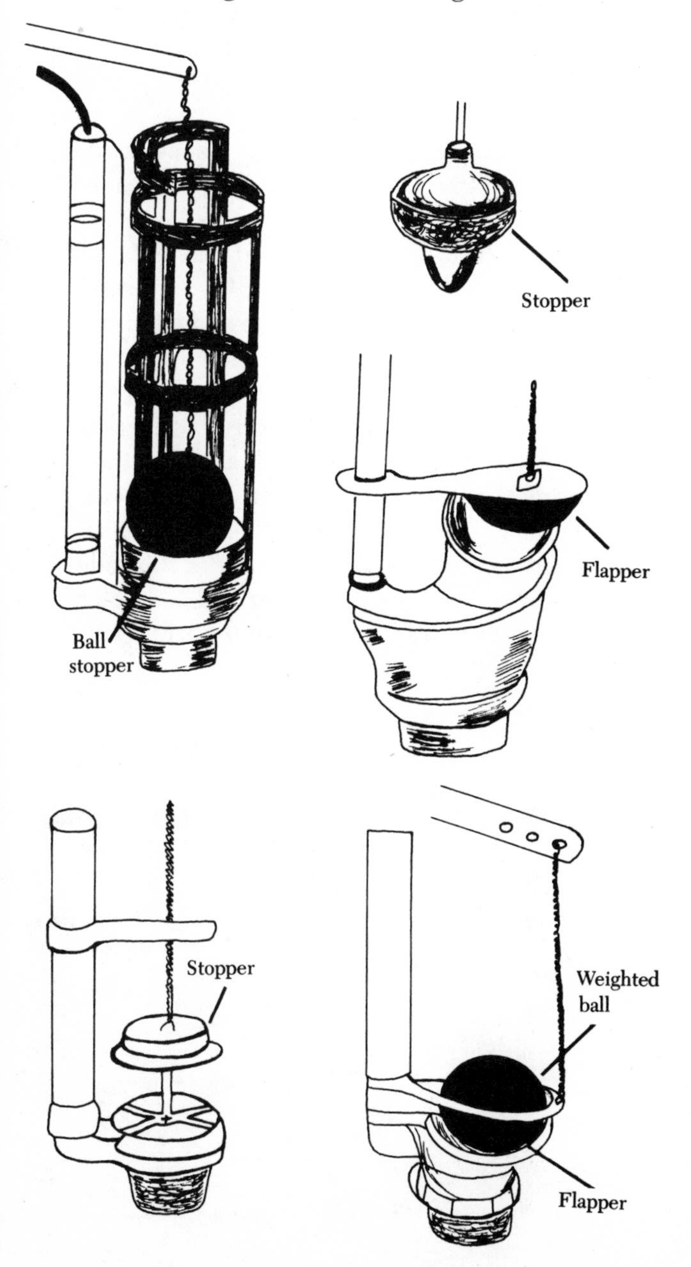

Some of the many tank stopper designs

Tank Stoppers

The stopper can be a simple, flat rubber disk, or there may be a half-ball molded to its underside. In older toilets, it may even be a ball that rides up and down inside a cage. The stopper is raised and lowered by the trip lever, either with a system of guide rods or a chain.

Stopper-Lift Mechanisms

The trip lever is connected to either a chain or a pair of guide rods which are attached to the stopper ball or flap that opens and closes over the tank drain. The drain allows water in the tank to empty into the toilet bowl during the flushing operation. Whatever the lift mechanism, it is designed and positioned to raise and lower the stopper over the outlet with unerring accuracy so that the stopper always forms a watertight seal whenever it is closed.

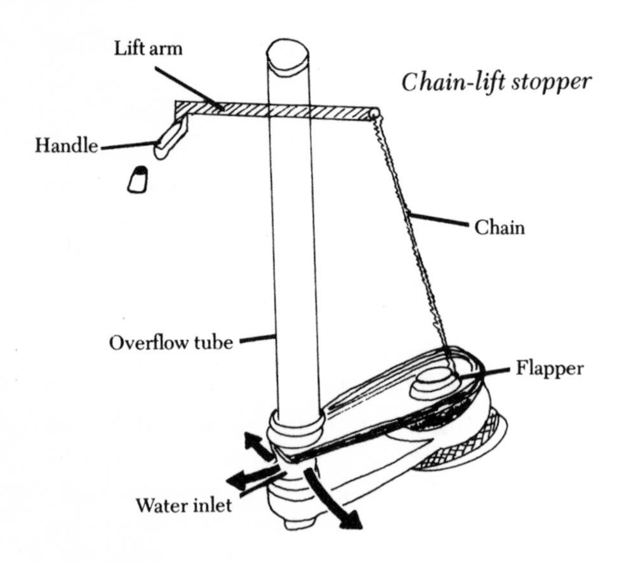

Chain Lifts

If your toilet has a chain, it is hooked into one of the three holes in the free end of the trip lever and to the rim of a rubber flap-type stopper. When the flush handle is moved, the trip lever lifts upward, pulling the stopper with it. The pressure of water rushing down the drain holds the stopper open until most of the water has drained out of the tank. The chain must be long enough to allow the stopper to close, but also short enough so that it does not snag on any other part of the flush mechanism and still pull the flapper open properly.

The only way to make sure the chain is functioning properly is to open the top of the tank and keep flushing the toilet while observing the stopper action. You may have to shorten the chain by moving the hook in the trip lever to another link, or move the hook to one of the other holes in the end of the trip lever.

Rod-Lift Assembly

The rod arrangement consists of a pair of guide rods: one rod is connected to the trip lever and has a ring in its free end; the other is attached to a stopper ball and slides through the ring. The ball stopper must drop straight down into the outlet seat, so that if the rods bend out of line, they must be bent back into position. To repair a rod-lift assembly:

1. Shut off the water supply.
2. Drain the tank, holding the rubber stopper ball up so that as much water as possible can drain out of the tank.
3. Examine the guide rod and the stopper ball for any wear or damage, and to make sure the rods are securely connected to the trip lever and the stopper.
4. If the rods have become disconnected, reconnect them. If they have been badly bent or appear weak, they should be replaced by a similar set (rods are sold at most large hardware stores). Or you could install a stopper-and-chain arrangement.
5. When you have either installed a new set or reconnected the old rods (or when you are simply examining the rods), verify that they are straight and plumb and that they bring the stopper ball directly into its seat. Do this by working the flush lever. You may need to bend one or the other of the rods slightly, or move the upper rod to one of the other holes in the end of the trip lever.

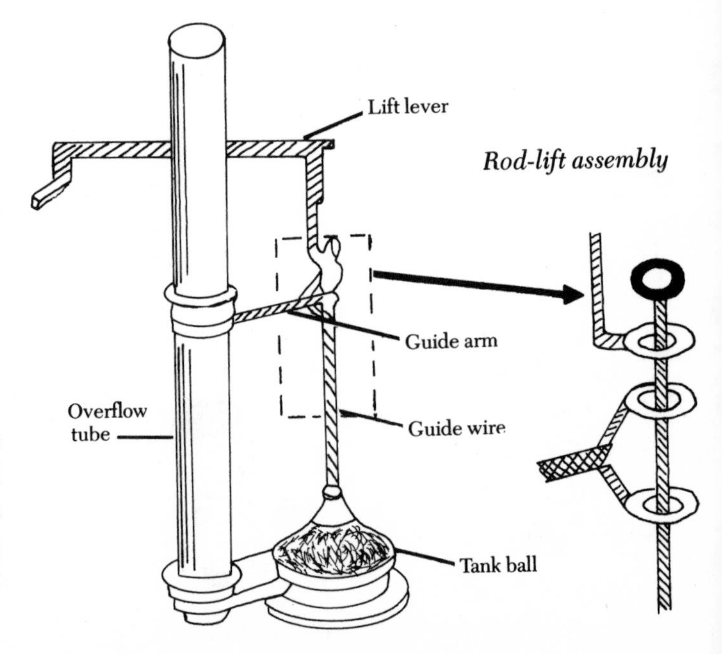

6. When the assembly is working correctly, replace the tank top and turn on the water supply.

Replacing Stoppers

There are, unfortunately, a surprising number of stopper designs and just as many ways of constructing the seat they fit into (or over). Consequently, you'll have to be careful about the type of stopper replacement you purchase. Some replacement units claim to fit most or all tanks. They do not. The port which the stopper must close off is connected to the base of the overflow pipe so that overflowing water can drain into the bowl of the toilet by bypassing the stopper. But the design of the pipe and the outlet can vary considerably. Some of the overflow pipes have raised bases, or they lock into a socket that is elevated above the rim of the outlet, with the result that stoppers which are meant to be clamped to the overflow pipe cannot be brought down close enough to the rim of the port to close it off effectively.

A good way to buy a replacement stopper is to take the old unit with you to a hardware or plumbing supply store and try to get an identical model, or at least one that appears to be attached to the overflow pipe in the same manner. The best method of buying a replacement stopper is to take the overflow tube and outlet with you, but that requires dismantling the tank and bowl (see page 49), and may be impractical.

There are a few stoppers sold that are attached to the drain rims. These can be glued with epoxy inside the outlet port, and operate separately from any hinging on the overflow pipe. This type of stopper will, indeed, fit most toilets, but it costs almost as much as an entire flush valve—around $5.00. To replace a stopper:

1. Shut off the water supply valve and drain the tank by flushing the toilet.
2. Disconnect the stopper from the trip lever, either by freeing the chain or the guide-lift rods.
3. The stopper is attached to the base of the overflow tube. It may have flaps protruding from it which have holes in them that fit around an eared flange bolted to the overflow pipe. But there are any number of other ways the stopper can be hinged to the pipe. Fortunately, all of them are simple enough to figure out.
4. Remove the stopper and examine it for wear or damage and replace it if it is in less than perfect condition.
5. Attach the replacement stopper to the overflow pipe, following the installation directions that accompany it.
6. Connect the stopper to the trip lever.
7. Test the stopper to be sure it opens and closes the outlet port properly.
8. Turn on the water supply and flush the toilet several times to verify that the stopper is working properly. If it does not close the port completely, you may have to realign it by twisting it to one side or the other, or by raising or lowering its hinge connection on the base of the overflow pipe.
9. Replace the top on the tank.

Overflow Pipes

The purpose of the overflow pipe is to siphon off water that rises too high in the tank. Invariably, the pipe is attached to the tank drain via the stopper valve seat. To remove a damaged overflow pipe from the tank, disconnect the stopper and rotate the pipe counterclockwise.

If your problem is leakage between the tank and bowl, or if you wish to replace both the overflow pipe and stopper seat, the tank must first be removed from the toilet bowl.

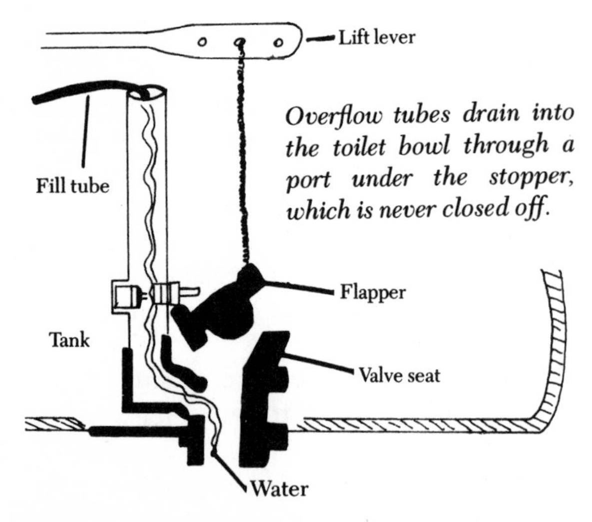

Overflow tubes drain into the toilet bowl through a port under the stopper, which is never closed off.

Replacing the Overflow Pipe and Stopper Seat

1. Shut off the water supply and flush the toilet.
2. Sponge out all water left in the bottom of the tank.
3. Disconnect the water supply line from the flush valve tailpipe.
4. Undo the nuts holding the bolts protruding from the bottom of the tank, through the back of the toilet bowl. There is one bolt on each side of the bowl.
5. Lift the tank off the bowl and turn it upside down. Place it carefully on the floor.
6. The spud nut around the threaded base of the seat valve may be metal or plastic. It is also bigger than you can grip with a 10″or 12″ pipe wrench, but you can free it with a large pair of channel-lock pliers, a pipe wrench, or a strap or chain wrench. Once you have found a tool that will grip the spud nut, rotate it counterclockwise and remove it.
7. The overflow pipe and valve seat can now be removed from the tank. Disengage the stopper from the trip lever and pull the pipe assembly out of the tank.
8. When installing a replacement overflow pipe assembly, be sure the proper washers are on the tailpipe of the valve seat, then push the tailpipe through the drain hole in the bottom of the tank. Place the spud washer against the bottom of the tailpipe and tighten the nut over it.
9. Place the tank on the back of the toilet bowl. The outlet seat should fit snugly into the drain hole at the back of the bowl, and the bolts in the bottom of the tank must go through their appropriate holes in the bowl.
10. Align the tank with the bowl, or with the wall behind it. There is a small amount of play in the bolt holes so you can adjust the tank to the proper alignment.
11. With the tank properly aligned, tighten the nuts on the tank bolts until they hold the tank snugly to the bowl. The nuts should be hand-tightened, then alternately tightened with a wrench to bring the tank down against the bowl evenly.
12. Connect the water supply line to the flush valve tailpipe.
13. Attach the stopper-lift mechanism to the trip lever.
14. Turn on the water supply and flush the toilet. Observe the base of the tank for leakage and be sure the tank stopper is working properly.
15. Replace the tank top.

Replacing a Toilet

Many of the new toilet units offer a better flushing action than the older ones. Besides being more efficient, newer models offer a considerable choice of design, ranging from units that hang off the wall to ultra-effective configurations that use a minimum of water. Still others

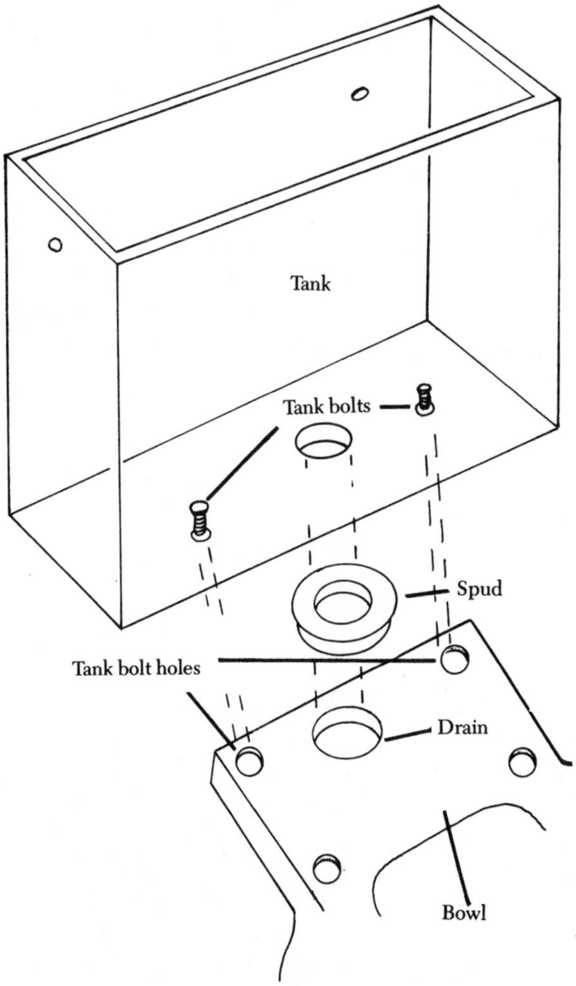

Toilet tanks are held to their bowls by a pair of large bolts.

are one-unit affairs with their tanks and bowls molded into a single piece of vitreous china. It does not much matter which design you select, they are all held in place in your bathroom by, believe it or not, two brass bolts, and you can remove your old unit and install a new one in about an hour's working time. The tools you will need include a 10″ pipe wrench, some sealing tape, a pair of pliers, and perhaps a putty knife.

Dismantling the Old Toilet

1. Shut off the water supply valve.
2. Flush the toilet and sponge out all water remaining in the tank.
3. Disconnect the water supply line from the tailpipe of the flush valve.
4. Using pliers or your wrench, undo the nuts holding the tank bolts. Some older toilets have an L-connection between the bottom of the tank and the back of the bowl. Undo the coupling nuts that hold the L. If the couplings are rusted, you can saw the L in half with a hacksaw.
5. Disengage the tank from the toilet bowl. Depending on the type of unit you have, lift the tank off the top of the bowl or undo the screws that hold it against the wall.
6. There are either two or four semicircular caps glued to the base of the toilet bowl. Using a putty knife or screwdriver, pry off the caps. You will find bolts and nuts in the base of the bowl.
7. Modern toilets use two bolts, one on each side of the bowl. Older models have two bolts plus two screws, or four bolts. Undo the bolts and/or screws.
8. The bowl was installed on a caulking of plaster of paris, grout, or plumber's putty around the perimeter of its base. Over the years the caulking may have become tenaciously hard, so you might have to chip away at it with your screwdriver. You can also wiggle the bowl from side to side to help break the seal.
9. When the bowl is loosened from the floor, pull it straight upward until it is clear of the bolts. There may be a little water in the bowl, so don't be surprised if your feet get splattered.
10. There is a ferrule plate in the floor which is anchored around the lead bend that leads to the soil stack hidden in the wall. You

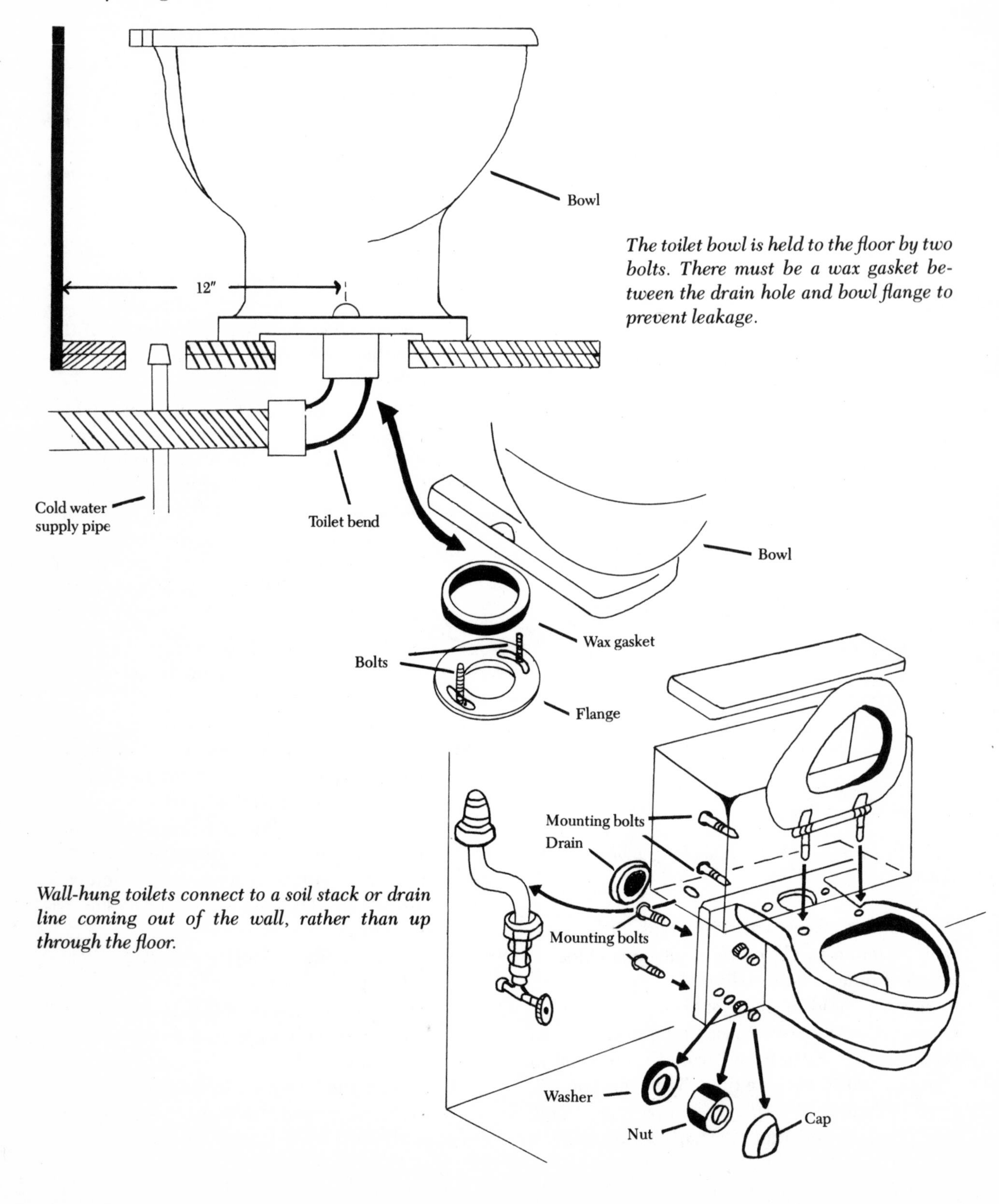

The toilet bowl is held to the floor by two bolts. There must be a wax gasket between the drain hole and bowl flange to prevent leakage.

Wall-hung toilets connect to a soil stack or drain line coming out of the wall, rather than up through the floor.

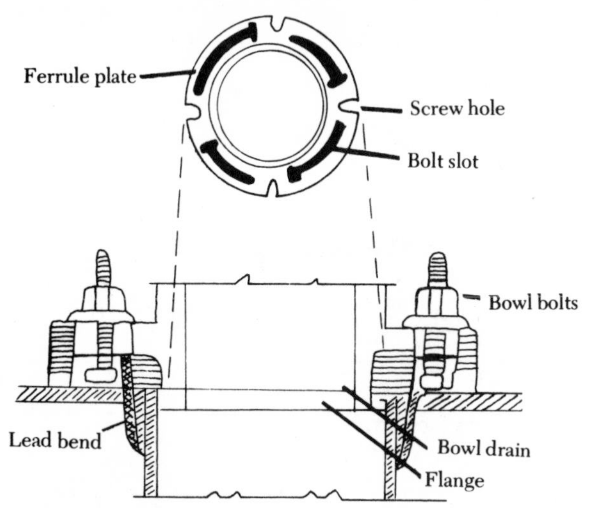

The ferrule plate is a metal disk with slots in it for anchor bolts.

probably will not recognize the plate at first because it is covered with a mound of gooey, dirty wax. Scrape the wax off the plate with your putty knife—it should come off easily. The ferrule plate is secured to the lead bend and has keyhole-shaped slots which hold the bolts upright. Slide the bolts out of their slots and thoroughly clean the ferrule plate. Then clean the floor around it.

11. When the ferrule plate and the immediate area around it is completely clean, inspect the lead bend closely for any signs of cracks or holes. If you find any damage, it should be immediately repaired by soldering the crack or holes.

12. Discard the old toilet tank and bowl. If you are replacing a wall-hung toilet with a unit that has a smaller tank or one that does not have to be attached to the wall, you may have to make some repairs to the wall. Now is the time to patch any screw holes and paint the wall, if you wish.

INSTALLING THE NEW TOILET

Usually, the distance from the finished wall to the rear of the ferrule bolts is 12", which allows a new toilet to be installed over an existing ferrule plate. Before you begin, you may discover that your new unit will not touch the wall after it is installed—even though the slots in the ferrule plate give you an inch or two of leeway. It's possible to move the new toilet back against the wall if you wish, but be warned that repositioning the ferrule will probably require some major renovation of the lead bend: you will have to open up the floor, use a torch to melt the lead connections, reposition the bend and the ferrule plate, then close up the floor again. If you feel you must position the toilet flush against the wall, the most efficient way of doing the job is to call in a plumber with the proper equipment. Or

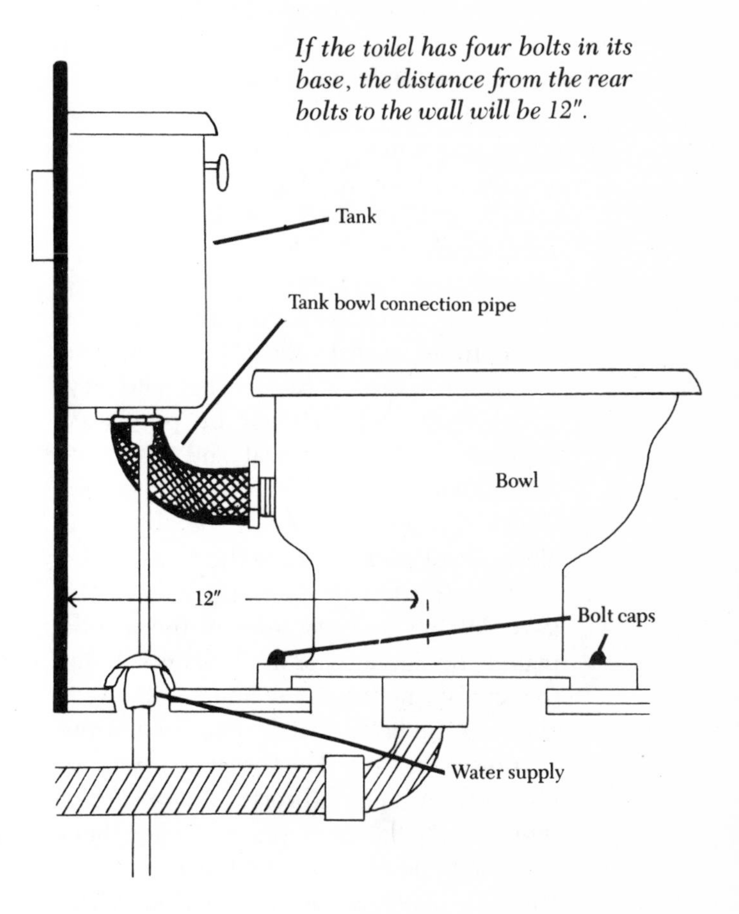

If the toilel has four bolts in its base, the distance from the rear bolts to the wall will be 12".

you can remove the lead bend altogether and use a plastic connection.

The procedure for installing a new toilet is as follows:

1. Check the new unit for any signs of damage and be sure you have all of the necessary components to assemble it. The tank should have a flush valve in it that is properly secured, and there should be enough nuts, bolts, and bolt caps for the base of the bowl. In addition to the toilet itself, you will need a wax ferrule gasket and new ferrule bolts, which must be purchased separately. The toilet seat and cover are also individual purchases.
2. Check the ferrule plate to be sure it is clean, level, and secure to the lead bend.
3. Position the ferrule bolts upright in the keyhole slots on both sides of the ferrule plate. The flat head of the bolts fits under the groove in the wide end of the slot, allowing the bolts to slide into the narrow portion of the slots. The threaded shanks of the bolts must extend vertically upward; if they seem to flop over, you can brace them with a mound of plumber's putty.
4. Turn the toilet bowl upside down and place the new wax gasket around the outlet at the bottom of the bowl. Press it gently into place, making certain that it is flush against the bowl.

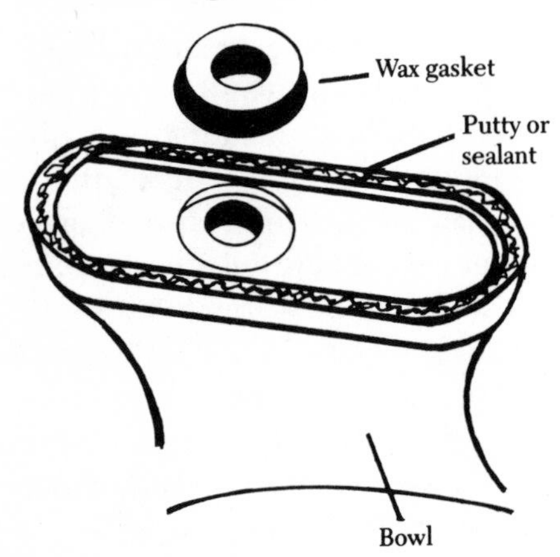

The wax gasket is pressed to the bottom of the bowl and surrounds the drain hole.

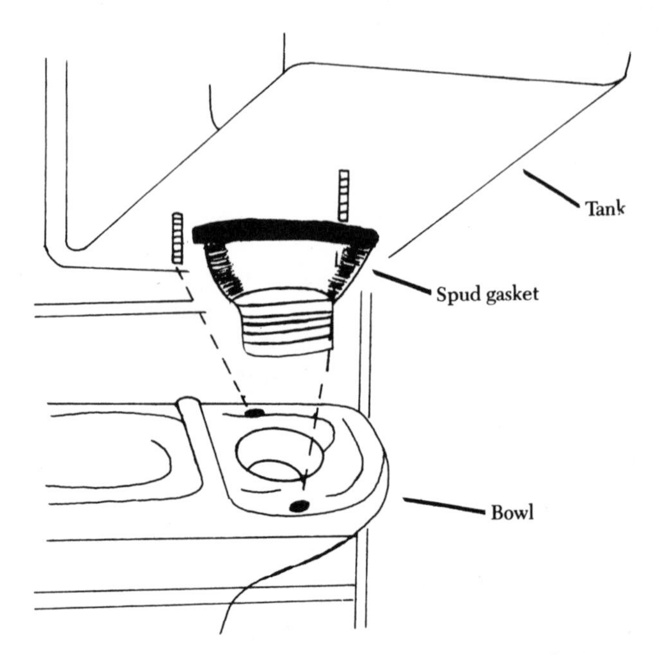

The toilet tank has a spud gasket that fits into a drain hole in the back of the bowl.

5. Lower the bowl down over the ferrule plate. You may have to move the bolts slightly to get them to extend through their holes in the base of the bowl. Press the bowl evenly down on the floor. You'll have some leeway so that you can twist the bowl enough to align it with the wall.
6. After the bowl has been properly aligned, place the appropriate washers on the ferrule bolts and hand-tighten the nuts to fasten them. Finish by alternately tightening the nuts with a wrench until the bowl is secure and does not rock.
7. Place the tank on the back of the bowl so that its bolts protrude down through the back of the bowl.
8. Hand-tighten the nuts to the tank bolts, then alternately tighten them with a wrench.

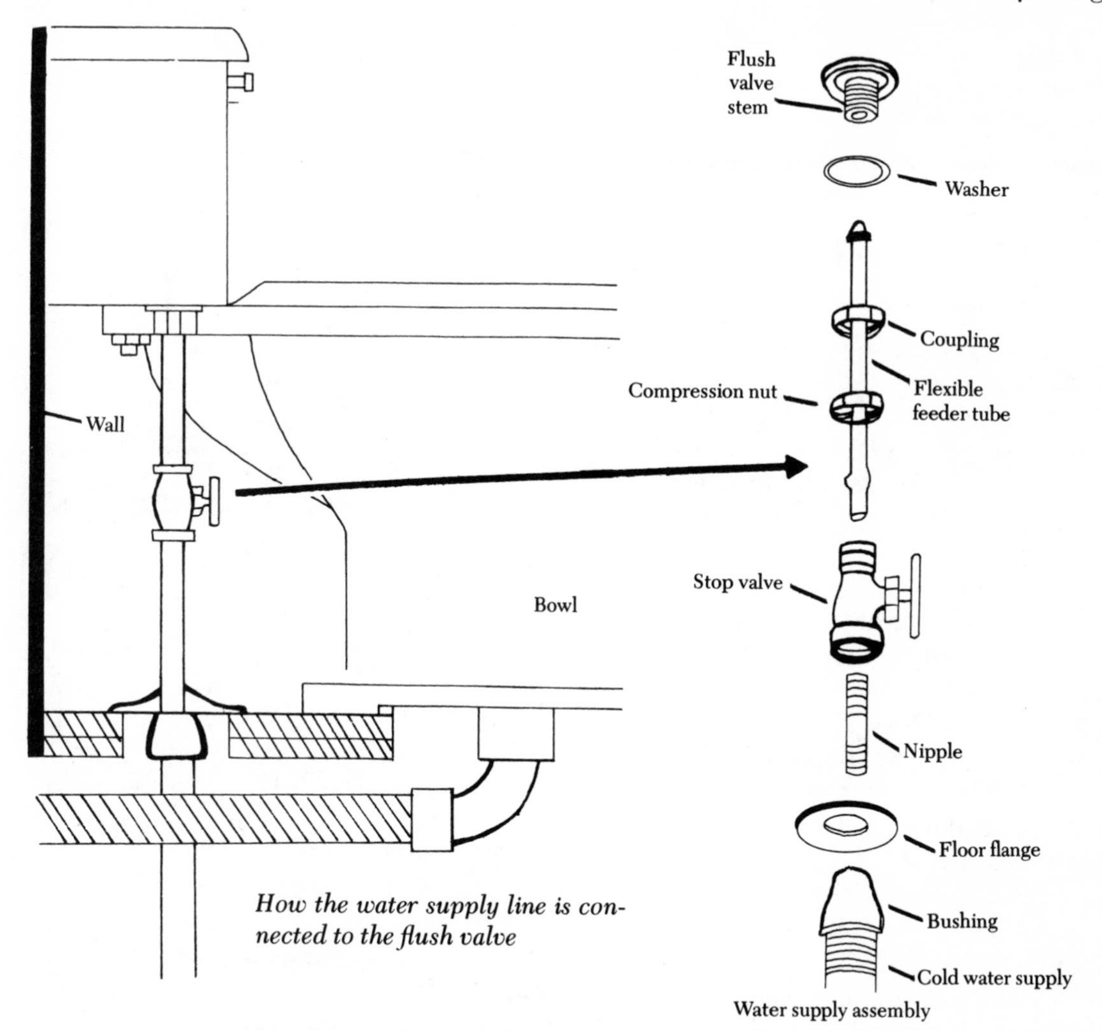

How the water supply line is connected to the flush valve

9. Connect the water supply line to the tailpipe of the flush valve.
10. Turn on the water supply and check your connections while the tank is filling. Tighten the coupling to the water supply line if there is any leakage. Check the water level in the tank and adjust the flush valve so that the water stops approximately ¾″ below the rim of the overflow pipe.
11. Flush the toilet. Inspect the base of the bowl closely to be sure there is no leakage at the ferrule gasket. If there is a leakage, try tightening the ferrule gasket bolts. If that doesn't work, you will have to get a new wax gasket, dismantle the toilet, and start all over again. Fortunately, the chances are that there will be no leak. Be sure to check the connection between the bowl and the tank for leakage there, also. If there is any sign of water, tighten the tank bolts.
12. Squeeze grout, caulking, or plaster of paris into the joint around the base of the toilet bowl and smooth it out to fill the gap between the base and the floor.
13. Glue the corresponding caps into place over the ferrule bolts. You may discover the bolts are too long for the caps to fit over

them; if so, saw off their ends with a hacksaw.

14. Put the tank cover on the tank.

Toilet Seats

You may have to replace a damaged toilet seat and its cover. Or you may make the change purely for decorative reasons. You can purchase plastic, wooden, and even cushioned toilet seats in a host of colors and designs, and installing them on any toilet is one of the easiest of all plumbing repairs.

1. Close the seat and cover.
2. Undo the nuts on the seat bolts. The bolts extend from the seat-and-cover hinge post through the top of the bowl; the nuts are on the underside of the back of the bowl.
3. Remove the washers from the bolts and lift the seat and cover out of the bolt holes.
4. Clean the top and undersides of the bowl around the bolt holes.
5. Place a washer on each of the bolts of the new unit and insert the bolts through the proper holes.
6. Place a washer on each bolt and hand-tighten their nuts; then finish tightening them with an open-end wrench.
7. Open and close the seat and its cover to be sure they are working properly.

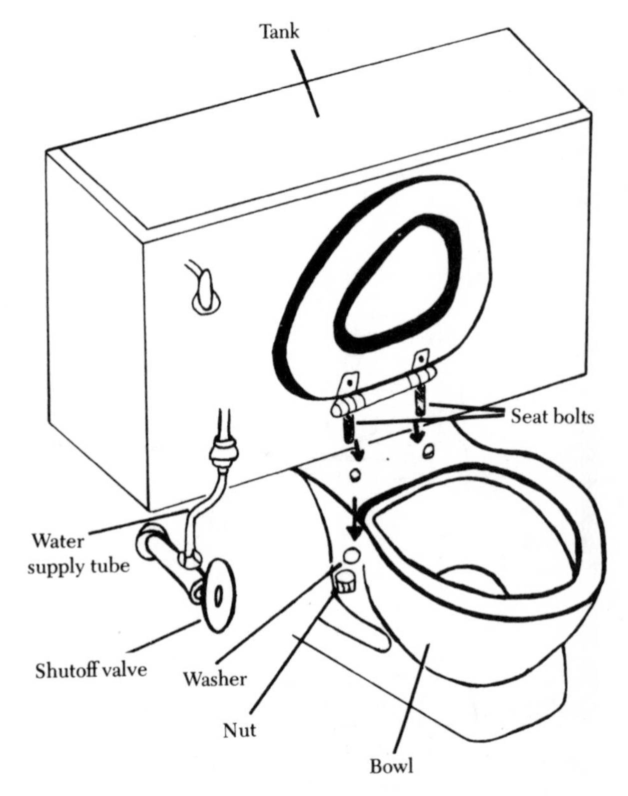

How a toilet seat is attached to the bowl

CHAPTER 4

Improving Sinks, Showers, and Bathtubs

Sinks

In technical terms, sinks are what you use in the kitchen; basins are sinks that are somewhat deeper, and are found in basements and laundry rooms; lavatories are the sinks in bathrooms. In the long run, they all amount to the same thing: a sink. Each is a receptacle for holding water and each has a well in its bottom that contains a drain, which in turn is attached to a trap and then to the house drain-waste-vent system (DWV). Sometimes a sink will have holes bored through the back rim of the unit to contain its faucets, and bathroom sinks always have an overflow port so that water rising above a given level will automatically drain out of the sink.

Sinks, in all their forms, hardly ever develop any disorders once they have been installed. They rarely crack or break or leak, but if they do they must be replaced. Of course, sinks do become outmoded: the decor in a kitchen or bathroom may change, the porcelain may become chipped or uncleanably stained with old age, or you may simply decide to replace the unit with a more modern one.

The different kinds of sinks are currently manufactured from a variety of very durable materials, including plastic, fiberglass, steel, iron, porcelain, and aluminium. They all work equally well, and they are all installed in approximately the same manner. Depending on its specific design, a sink can be hung from a metal bracket attached to the wall with screws. It can be stood on legs, or supported by a counter which must have an opening cut out of its surface. Once the sink is in position, its attending faucets must be connected to the hot and cold water supply lines. Then the drain must be attached to the house drain system with drainpipe that is either 1¼″ or 1½″ in diameter. Replacing or installing a new sink is neither difficult nor very time-consuming.

Replacing Sinks

Before you install a new sink or replace an old

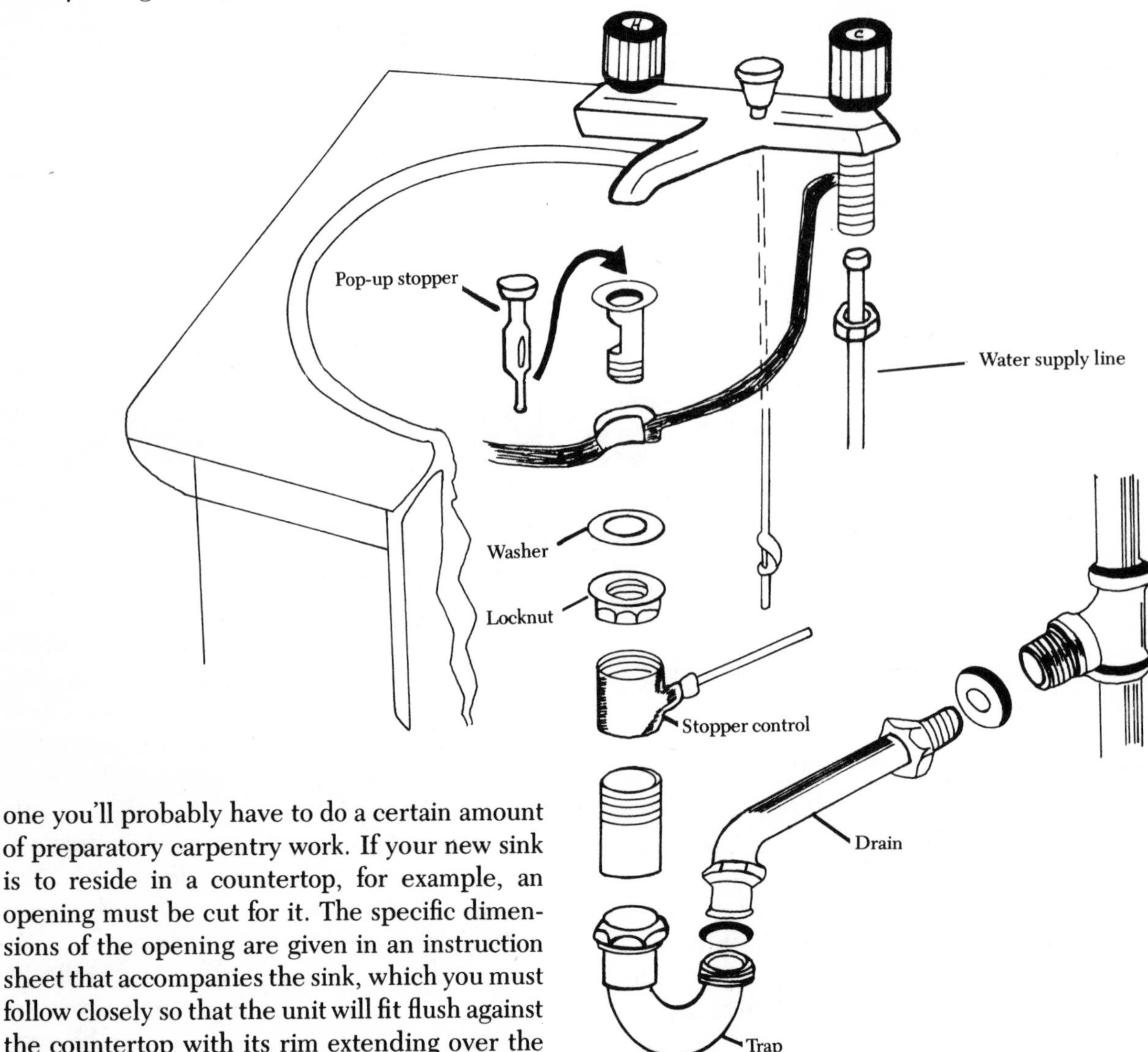

The assembly of pipes under a typical sink is not as confusing as it looks.

one you'll probably have to do a certain amount of preparatory carpentry work. If your new sink is to reside in a countertop, for example, an opening must be cut for it. The specific dimensions of the opening are given in an instruction sheet that accompanies the sink, which you must follow closely so that the unit will fit flush against the countertop with its rim extending over the opening. If the sink is to stand on either two or four legs, it must be attached to the wall in some manner. If the unit merely hangs from the wall, it must fit over a metal bracket that has to be positioned on the wall and attached to wall studs or a wooden brace. All of the procedures for positioning a replacement sink solidly are detailed for you in the installation information provided by the sink manufacturer.

Once you have removed the old unit, you will have bared a section of the wall that may not have been painted in years. Any renovation to the walls or floor around the sink are most easily done before you install the replacement unit.

No matter how a sink is installed, the end result must be a stable, level fixture. This means that the sink should not slide or wobble, or move in any way. If you lay a carpenter's level across the rim of the sink, it should read level in two directions, from front to back and from side to side.

The general procedure for replacing any sink and hooking it to the house plumbing systems is as follows. Each unit will differ slightly, however, so be sure to follow the manufacturer's instructions closely.

1. Shut off both the hot and cold water supply valves that are connected to the old sink.
2. Disconnect the water supply lines to the faucets. If you can reach them easily, undo the nut between the supply line and the tailpipe of each faucet. If you do not have a basin wrench and cannot reach under the sink to the faucet tailpipes, undo the nuts that hold the supply tubes to the top of the shutoff valves. As long as the valves are closed there should be little or no water leakage when the supply lines are removed.

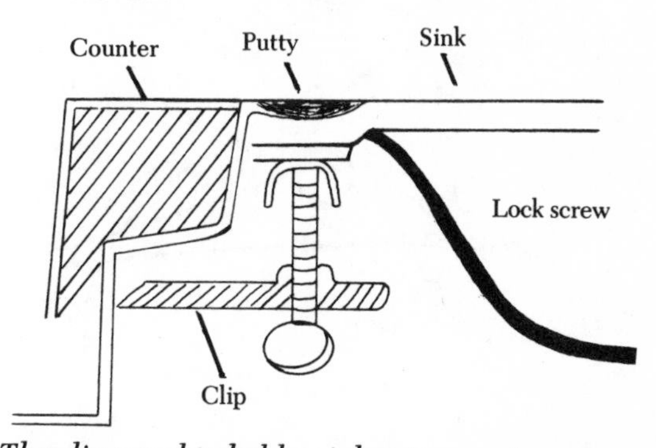

The clips used to hold a sink to its counter may vary in design, but they all rely on a threaded screw or bolt to secure the fixture in place.

Anatomy of a sink. Traps can be removed from the drain line for cleaning.

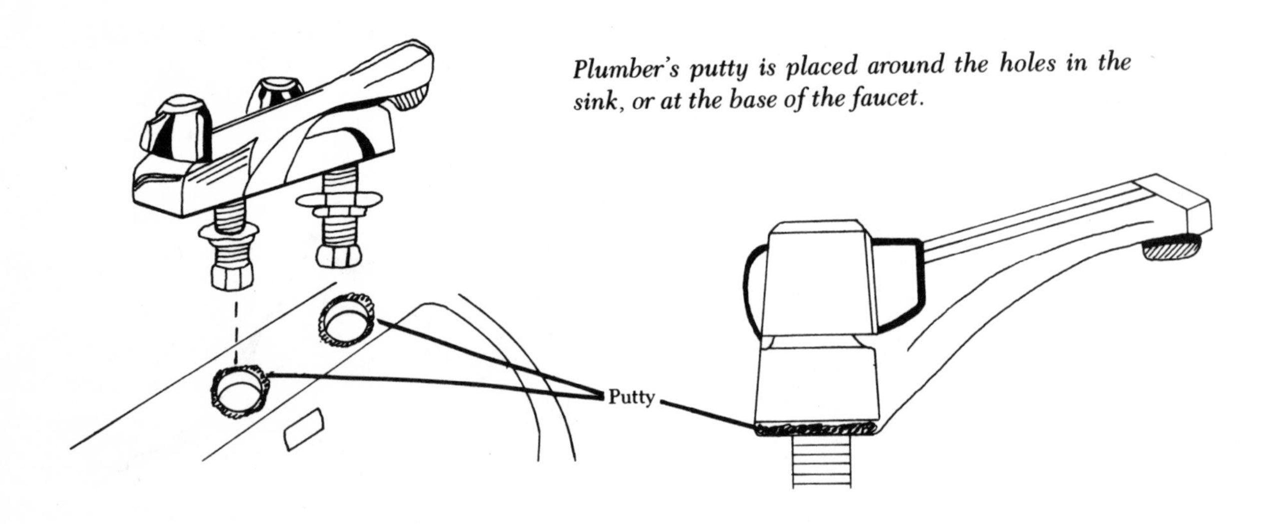

Plumber's putty is placed around the holes in the sink, or at the base of the faucet.

3. Loosen the coupling nut that holds the drain trap to the drain tailpipe.
4. If the sink resides in a countertop, it is held in place by a series of clips that are each locked to the underside of the sink rim by a set screw. Each of these clips must be loosened with a screwdriver or, in some instances, a pair of pliers. The sink can then be pushed up through its opening. With a sink that hangs from the wall, free all of its plumbing connections and then slide the unit upward until it comes out of the wall bracket that holds it in place. If the sink is freestanding on its own legs, it can be moved as soon as the plumbing lines are all disconnected.
5. Complete whatever carpentry work is necessary to the wall and floor. If the new sink will not be positioned exactly as the old sink was, you may also have to change the water supply lines and the drainpipe itself (see Chapter 5 for information on working with pipes). You can purchase flexible copper or plastic tubing that easily makes the connections between existing shutoff valves and the faucets in a new sink. Plastic drain and trap kits can be found in any hardware store or plumbing supply outlet. These kits include all of the pipes, trap configurations, washers, and couplings necessary to accomplish whatever connections you have to make with the drain line.
6. Following the manufacturer's instructions, assemble the faucets to the new sink before you install the unit. Faucets are normally purchased separately from the sink (they can cost more than the sink itself), and come with all of the washers and nuts needed for installation. It is necessary to surround the faucet openings in the sink rim with plumber's putty. Roll a lump of putty between your hands until you have a string about a quarter of an inch thick, and wrap it around the edge of the hole. Insert the faucet through the hole and press it down against the rim. Then attach the washer and nut to the faucet tailpipe and hand-tighten the nut underneath the sink

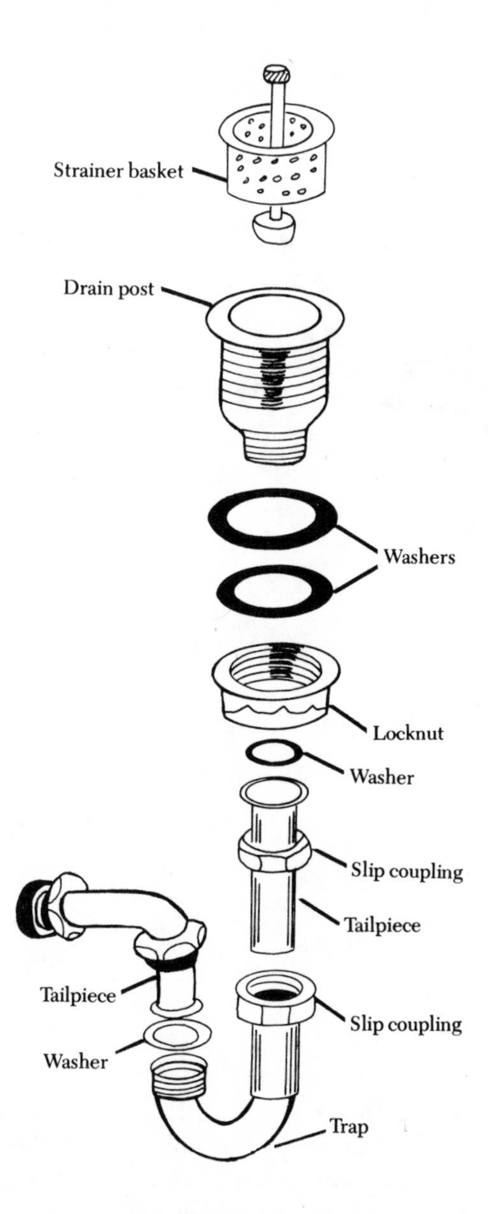

Anatomy of a drain basket, drain line, trap, and waste line

rim. Finish tightening the nut with a wrench until the faucet is secure. Be careful not to overtighten the nut, or you may strip the threads on the faucet tailpipe or crack the sink (if it is made of porcelain or some of the plastics).

7. If the faucet tailpipes will be close to the wall and difficult to reach, you might consider attaching the flexible water supply tube to them before you install the sink, so that your final connection will be at the more accessible shutoff valves. Simply wrap plastic pipe-sealing tape around the "male" threads of the faucet tailpipe, and tighten the connector nut on the end of the flexible tube to the tailpipe.
8. The drain may or may not have been purchased as a separate item. Again, place a roll of plumber's putty around the rim of the drain hole, then press the drain basket through the hole and tighten the locknut and washer against the bottom of the sink.
9. Position the sink and secure it so that it will not move and is level in all directions.
10. Wrap the drain tailpiece with plastic pipe-sealing tape (or coat the threads with plumber's putty) and screw it into the bottom of the drain.
11. Slide a rubber gasket over the tailpiece and tighten the tailpiece to the drain with its lockwasher.
12. If the sink you are installing has a pop-up drain assembly, the stopper-valve section is attached to the bottom of the drain tailpipe with a slip coupling that must be sealed with plastic tape or putty. Be sure to position the rod in the stopper valve at the proper angle so that it can connect to the drain plunger rod that comes down between the faucets. The two rods are connected by a spring metal clip. After you have assembled the rods, work the drain plunger several times to be sure it opens widely enough and closes tightly. If it does not function properly, adjust the clip on the connecting rod.
13. The trap is attached to the bottom of the stopper valve with a coupling nut. Wrap sealing tape or putty around the threads of

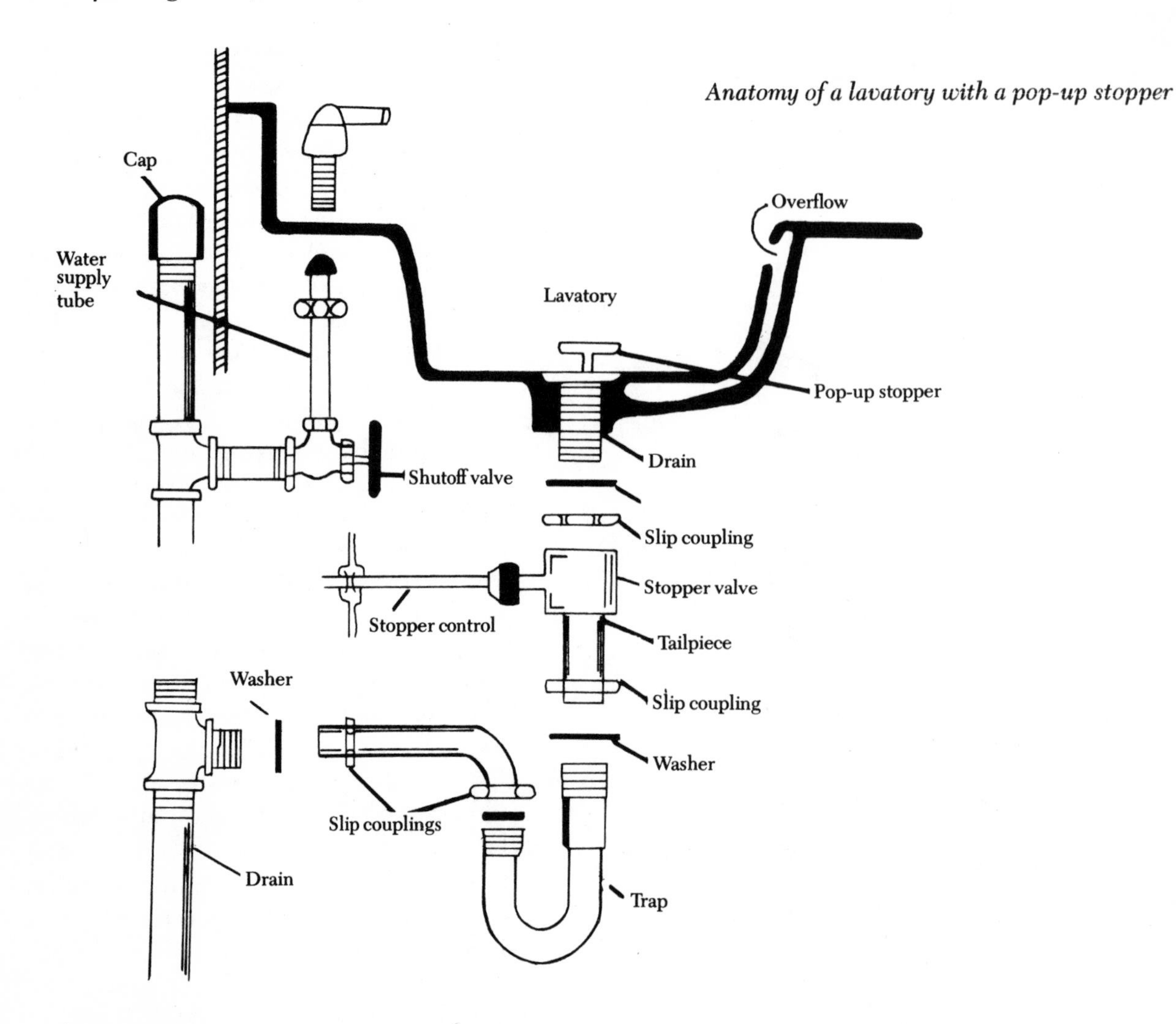

Anatomy of a lavatory with a pop-up stopper

the tailpiece (or the stopper valve, if one is being used), and connect the trap with its proper washer and a coupling nut. The nut can be hand-tightened and then given a quarter turn with your pipe wrench.

14. The free end of the trap also has a washer and connects to the waste line with a slip coupling. Again, seal the connection with plastic tape or plumber's putty.

15. The drainpipe is attached to the waste pipe coming out of the wall or up from the floor with a washer and coupling and pipe-sealing tape. So after you have made all the connections, the sink is attached to the house drain system.

16. Connect the supply lines to the shutoff valves or to the tailpipes on the faucets. The nuts must have sealers and be tightened with a wrench.

17. When all drain and supply line connections have been made, open the hot and cold water supply shutoff valves and turn on the faucets. Observe all of your connections for any signs of leakage. Wherever water appears, tighten the appropriate nut until the leakage stops.

18. Scrape off all excess putty around the drain, faucets, and the rim of the sink.

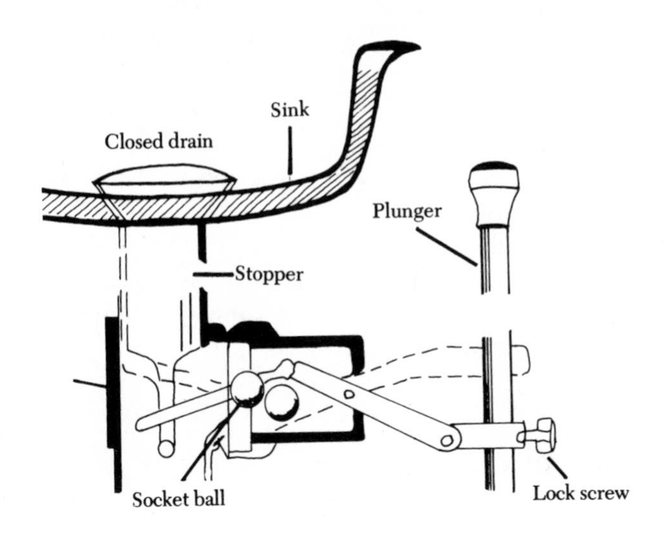

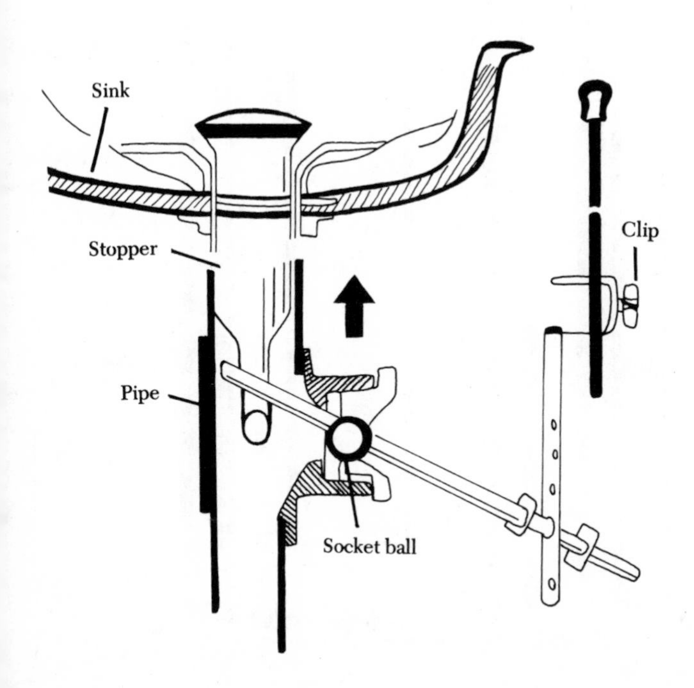

Two of the several ways pop-up stoppers are assembled

Showers

Showers can use either the water supply system in an existing bathtub or a separate supply-and-drain system that enters and leaves a cubicle known as a shower stall. You can build the stall by constructing four walls, one of which has a doorway, with studs and plywood sheathing. The sheathing is then covered with a metal lathe which supports a coat of cement that in turn is used as a base for ceramic tiles. The base of the stall can be either stone or plastic, or else made of wood, cement, and tiles.

An alternative to building your own shower is to buy a prefabricated fiberglass, plastic, or metal stall. These cost between $100 and $300 and come complete with walls, a plastic or stone shower base, faucets, and a showerhead—all of which can be assembled in a few hours and then connected to the house plumbing system.

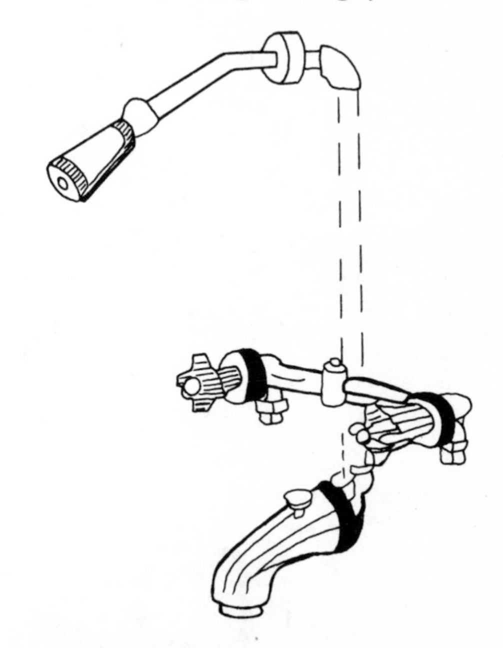

The king pin in a tub-shower faucet arrangement is the compression faucet fitting.

A third possibility arises when you have an existing bathtub that you want to use as a base for the shower, in which case you may have to cover the walls around the tub with a waterproof plastic or fiberglass tub surround (sold in kits), or with ceramic tiles. You will also have to extend the water supply system upward from the existing faucets to a mixer valve or another set of faucets, and then up to the showerhead.

Whether you are putting in a packaged shower stall or extending the existing lines behind a bathtub, the plumbing is about the same as far as the water supply system is concerned. But with the installation of a stall you must also make some drainage connections using 1½″ or 2″ drainpipe.

It is much easier to do the supply line work on a stall you are building than it is to work inside the walls behind an existing tub. The prefabricated stall kits include a compression valve which holds the faucets and the pipe leading to the showerhead. This valve can be used as a substitute for a single-handle valve or a thermostatic mixing shower valve (thermostatic mixers are expensive because they contain a device which holds the water at a desired temperature). Any of the three types of control valves can be installed in an existing water supply system behind a bathtub.

Plumbing Tiled Shower Stalls

The plumbing lines that service a tiled shower stall should be brought into the stall before the

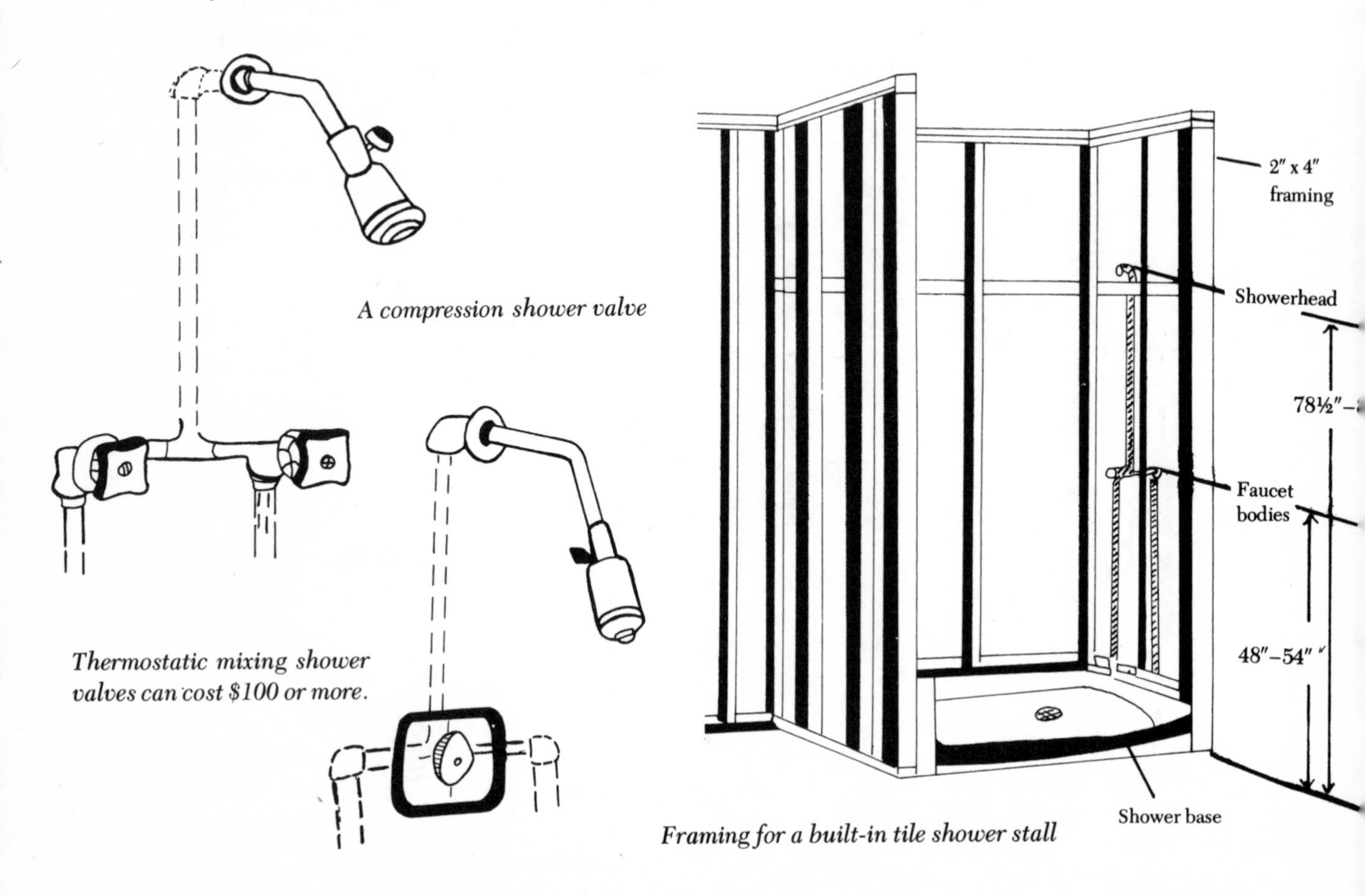

A compression shower valve

Thermostatic mixing shower valves can cost $100 or more.

Framing for a built-in tile shower stall

walls are closed up and tiled. The ½″ or ¾″ hot and cold water supply lines can be hooked up at any convenient point in the supply main by inserting a T-connection in the main and then running the lines to the stall.

Faucet handles are normally positioned between 32″ and 36″ above the floor and are linked by a compression-faucet fitting. The fitting accepts the hot water supply line under the left faucet and the cold water line under its right side. The showerhead pipe extends vertically up the inside of the wall from the center of the fitting. Showerheads are positioned between 5 and 6 feet above the floor and are connected to the supply pipe via a 90° elbow at the top of the pipe.

Two 1″ × 4″ boards should be notched in the facing edges of the studs on either side of the piping to secure the supply lines leading to the faucets, and also the top of the shower pipe. Only after you have made and tested all of your connections should the face of the stall be closed and tiled.

The base of tiled shower stalls can be a preformed fiberglass or stone pan, but these are manufactured in standard 30″, 32″, or 36″ squares that may not meet the specifications of the stall you are building. The pans are constructed with a 2″ drain hole in their center. A strainer is caulked to the outlet and attaches to a 2″ P-trap fitted directly under the pan.

If you are constructing your own stall flooring, select a floor drain with an integral P-trap and a drainpipe that is at least 2″ in diameter. The drain must be installed at the lowest point in the floor, and the drainpipe is run to the nearest soil stack (see chapter 6 for specific information on drainpipe connections). The floor of the stall should have a slight slope toward the drain that needs to be no more than ⅛″ per foot. Once the drain line has been positioned and the drain strainer installed, you can complete the cementing and tiling of the stall floor. The drain line can be galvanized steel, which is in keeping with

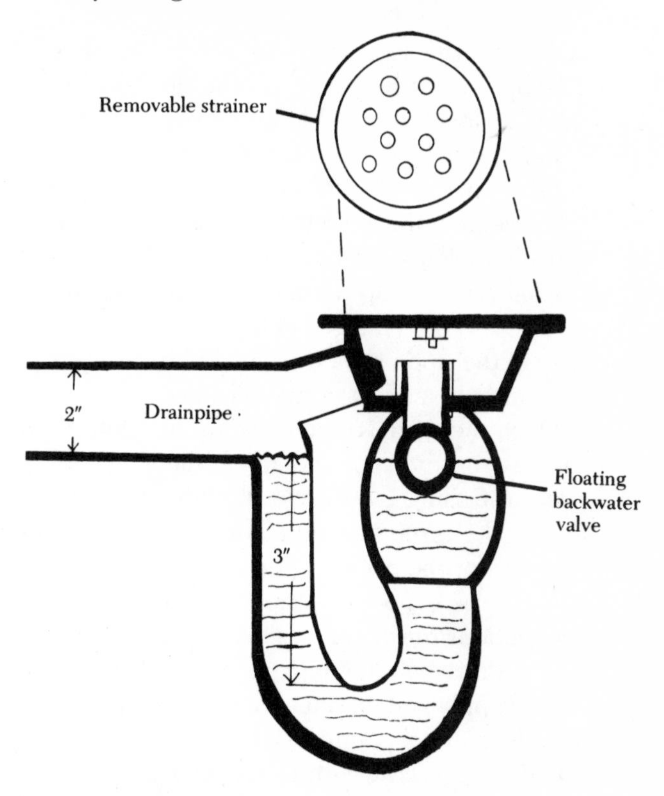

Cross-section of a floor drain used under a shower stall pan

most house drain systems; brass, which is extremely expensive; or plastic piping, which is not only more durable than the metals, but much easier to work with.

Assembling Prefabricated Shower Stalls

Prefabricated shower stalls are made of fiberglass or sheet metal and are sold in kits that include a stone or fiberglass base, a floor drain, four sides, assembly screws, faucets, a compression fitting, a showerhead, and a soap tray. The kits do not include the piping needed for either the supply lines or the drain, and the fittings needed to connect the stall to the house plumbing system; these must be purchased separately. Assembly of the prefab stall kits is fully described in the accompanying instruction sheets. The base must be placed on the floor and then made as level as possible by inserting blocks of wood under it. Once the base is level, nail the shims in place; then stand the base on one edge and assemble the drain basket and trap. The trap must have a grommet positioned between it and the underside of the drain hole in the base, and is screwed to the tailpiece of the drain. The waste port in the drain should be angled to accept the 2″ drainpipe leading to the soil stack.

Once the drain is connected to the trap, place the base on the floor and attach the sides of the stall to it using the screws provided in the kit. One of the three sides has holes drilled through it to accept the compression fitting and faucets. You can use copper or plastic pipe to make your connections to the water supply line, and these connections can be tapped anywhere in the water supply system, including behind the shut-off valves under a sink (see Chapter 6 for information on working with supply lines).

When you have attached the faucets and showerhead to the back side of the stall and completed assembly of the stall itself, stand the base on its shims and secure it in position. Then make your final hookups with the water supply and drain lines.

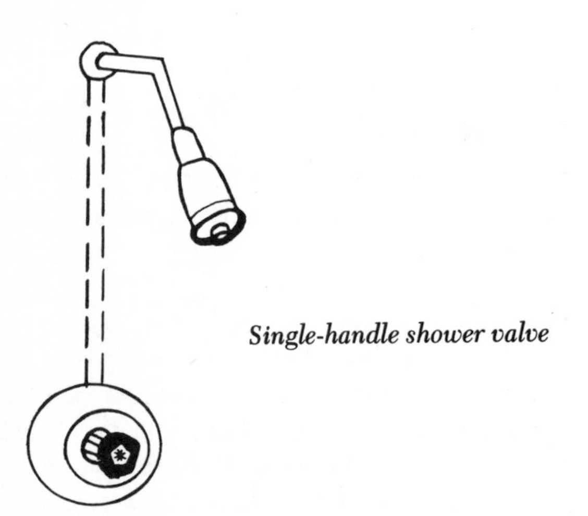

Single-handle shower valve

Making a Tub Shower

In order to get at the water supply lines that are behind a bathtub you may find you have to break through the face of the wall that houses the faucets. Sometimes there is access to the tub supply lines through a panel in the back side of the wall—that is, in the space behind the tub. Just as often the supply lines have been sealed inside the wall, which may well have been faced with tiles around the tub. If you must choose between breaking through a tiled wall or the wallboard on its reverse side, remember that it is easier to replace the wallboard than the tiles. Whichever you choose, enough of the wall around the pipes must be removed so that you can work. You may find that this requires making a good-sized hole around the faucets, as well as another hole 5 or 6 feet up the wall for the showerhead.

ADDING A SHOWERHEAD AND DIVERTER SPOUT

1. Shut off the hot and cold water supply valves and open the faucets to drain all water remaining in the pipes.
2. Remove the faucet handles and escutcheons (see page 30).
3. Rotate the spout counterclockwise to unscrew it from its supply pipe.
4. Remove as much of the wall around the valve as is necessary to give you plenty of space to work. You may have to go partway up the wall if you are adding a separate compression valve and faucets for the shower, or if you are replacing the existing compression valve with a single-handle fitting.
5. When you have exposed the compression fitting, examine the T-joint at its center. If the fitting was used only for a tub, the top part of the T may have a cap screwed into it

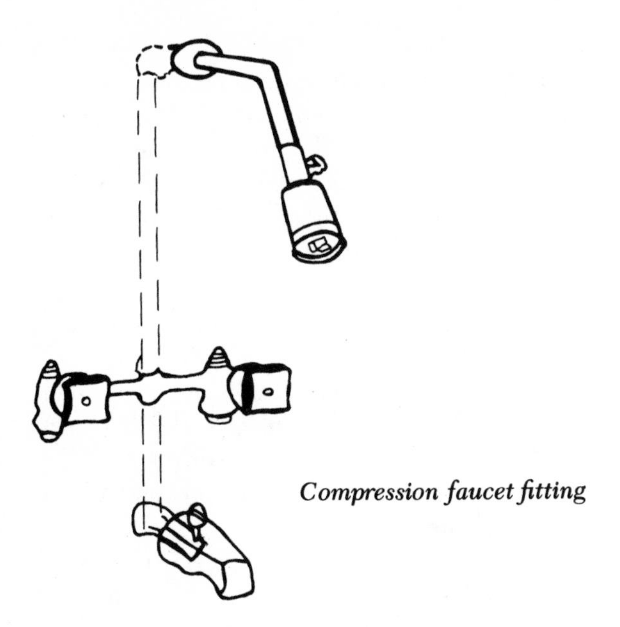

Compression faucet fitting

which can be replaced by a length of pipe that leads up to the showerhead. If you are content with the faucet arrangement as it exists, all you need to buy is a piece of brass or galvanized steel pipe, an elbow, a showerhead, and a tub spout containing a diverter knob that can be raised to divert water up to the shower. Also, buy a piece of pipe that is the proper diameter to fit in the compression-T socket (normally either ¾″ or ½″ in diameter), and is long enough to bring the showerhead to about 5 feet above the rim of the tub.

6. Assemble an elbow to one end of the pipe and push the pipe up the inside of the wall. Then wrap its threaded end with plastic pipe-sealing tape and screw it into the compression-T socket. Make sure the face of the elbow faces out of the wall.
7. Thread the showerhead pipe (which is usually chrome-plated and bent at a 45° angle) into the elbow and connect the showerhead.
8. The diverter tub spout is threaded to the same nipple used for the old spout.

REPLACING COMPRESSION FITTINGS WITH A SINGLE-HANDLE VALVE

If you wish to replace the compression fitting with a single-handle faucet fitting, or add a separate set of faucets for the shower, the plumbing entails a little more work. When you have exposed the plumbing you will probably find copper or galvanized steel piping. If it is galvanized steel, the compression valve is connected with threaded parts; undo them with your pipe wrench. If the piping is copper, it has been sweat-soldered at all its joints; the joints must be melted with a propane torch (procedures for working with galvanized steel, brass, or copper pipe are detailed in Chapter 5).

1. Dismantle the compression fitting and remove it from the hot and cold water supply lines.

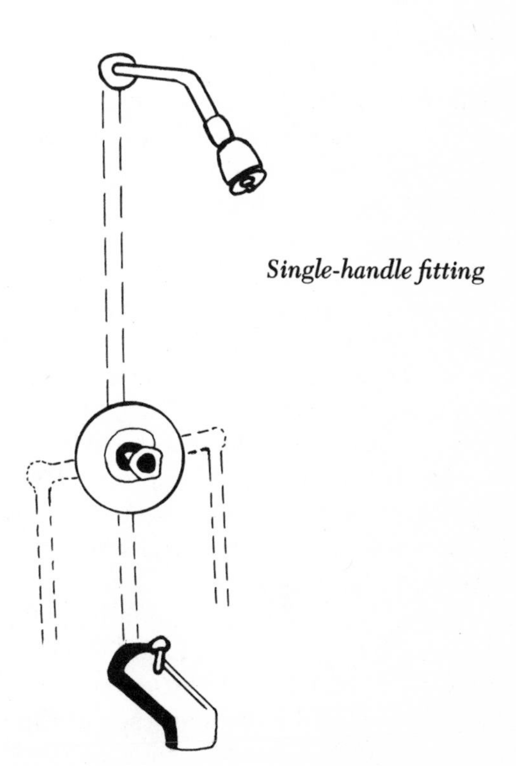

Single-handle fitting

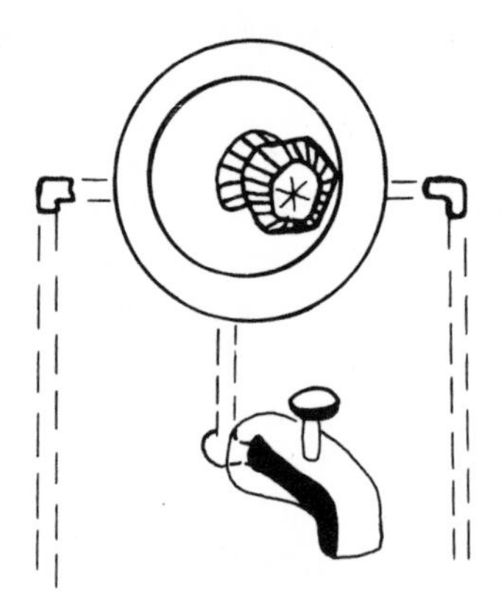

Single-handle bathtub fitting

2. Unscrew the tub spout and remove it.
3. Trace the position for the new diverter valve on the wall (it should reside approximately 36″ above the bottom of the tub), and knock a hole through the wall to accept the unit. The supply pipes will not be strong enough to hold the valve in place; it must be screwed either to a 1″ × 4″ support nailed across the face of two studs, or to a single stud, and should be located between the supply lines.
4. Position the valve in the wall, but do not secure it.
5. The supply lines must enter the sides of the unit, and the pipe leading to the showerhead must be connected to its top. Most valves have threaded sockets to accept either copper or plastic fittings, which in turn can be connected to copper or plastic piping. If you are assembling the unit to copper or plastic, the threaded fittings can be inserted and the appropriate pieces of pipe and elbows soldered in place *before* the valve is installed permanently in the wall. Each of the pieces of pipe to be assembled should be cut, positioned, and fitted beforehand to be sure they are the correct length. Basically, a pipe must extend from either side of the valve to an elbow which turns the pipe downward to meet the hot and cold supply lines. When you have assembled the nipples and elbows to the diverter valve, install the valve permanently in the wall, locating and securing it according to the manufacturer's instructions.
6. Complete the pipe connections between the valve and the water supply lines.
7. Measure and cut a length of pipe to run from the top of the valve to the center of the hole made for the showerhead.
8. Assemble a 90° elbow to one end of the showerhead pipe.
9. Connect the showerhead pipe to the diverter valve by sliding it up behind the wall and then fitting it into its socket in the valve. Solder or solvent-weld the pipe to the valve, making certain that the elbow is facing the wall.
10. Wrap plastic pipe-sealing tape or plumber's putty around the threads of the showerhead arm and screw it into the elbow.
11. Wrap the opposite end of the arm with tape or putty and attach the showerhead.
12. Secure the showerhead pipe to its 1″ × 4″ support with a metal strap.
13. Check the diverter to be certain the faucet body is in its correct position, and attach the handle to the faucet body.
14. A fourth pipe should run down from the bottom of the diverter to the spout position, ending with a 90° elbow which supports a short nipple that extends into the spout. The nipple must be threaded, either by soldering or solvent-welding a threaded fitting to its end; if the nipple is brass, it must have threads cut in its end. Measure and assemble the necessary pipes and fittings and connect them to the bottom of the diverter.

15. Screw the diverter spout to the end of the spout pipe.
16. Open the hot and cold water shutoff valves and test the entire assembly for leaks. If the threaded connections show signs of water, you will have to free the soldered side of the threaded fittings, tighten the fitting, and then resolder the connection. If a soldered connection leaks, simply re-solder it. If you are using plastic pipe and fittings and one of your solvent-weld con-nections leaks, the fitting must be cut out of the assembly and replaced. Only when you are absolutely certain there are no leaks anywhere in the assembly can you begin to consider closing up the wall.
17. Close the wall with pieces of drywall and replace the tiles, if there are any.

ADDING A SHOWERHEAD AND FAUCETS TO A TUB

It is often possible to add a second compres-sion faucet fitting for a showerhead, rather than a single-handle mixer valve, directly above the existing unit—provided that the compression fitting within the wall has a center T-joint that is plugged. Check for this when you are opening the wall to make working space.

Tub Surrounds

Anytime you elect to add a shower to an exist-ing tub, you must choose a way of protecting the walls around the tub and keeping water from splashing all over the bathroom. The splashing is easiest to contain simply by installing a rod and showercurtain or sliding glass doors across the side of the tub that faces the room. The curtain rod must extend along the full length of the tub, and should be positioned a few inches in front of the tub rim so that the curtain hangs at an angle into the tub, rather than straight down.

Protecting the walls around the tub is a more complicated task. The most common solution is to cement ceramic tile to the three walls, at least as high as the showerhead. A more modern solu-tion is to glue a tub surround to the walls. Tub surrounds can be purchased in kits that include three sheets of plastic or fiberglass, corner fram-ing, mastic, and a waterproof grout. The assem-bly and installation of the kits is explained in detailed manufacturer's instructions, which make the task seem simple enough. But there is one element to consider carefully before you spend $100 or so for a kit: it is likely that the walls surrounding your tub are neither plumb nor formed at a perfect right angle. It is critical that the three sides of the tub surround come to-gether in a watertight joint; if the existing walls are only slightly out of line, you will be able to overcome the discrepancies with extremely careful cutting and assembly of the surround pieces. But if the walls are badly out of line, or are in poor condition, the job may be easier and less time-consuming if you use ceramic tiles.

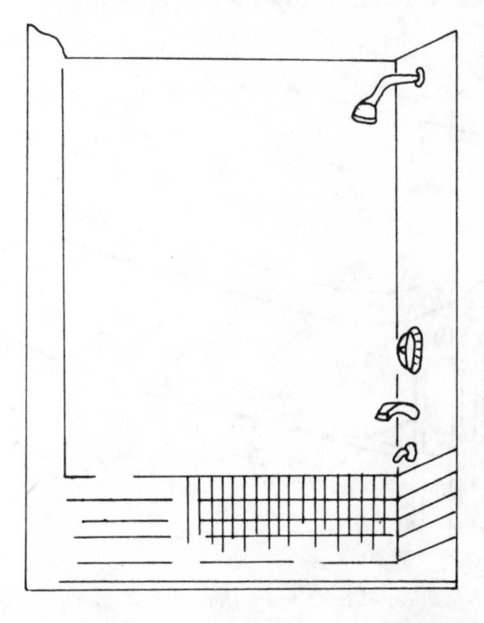

Tub surrounds are manufactured in masonite, plas-tic and fiberglass.

Bathtubs

Apart from occasionally unclogging their drains or stopper mechanisms, bathtubs rarely require any repairs. Thus, they are hardly ever replaced for any reason other than decorative purposes. Tubs are commonly manufactured from enameled cast iron, steel, or fiberglass. Because they are normally installed in a recess, they can demand a considerable amount of time when being replaced.

Replacing a Bathtub

If you are replacing an old-fashioned tub that stands on legs, all of the drain and supply lines and their fittings are probably out in the open, where they can be easily undone with a pipe wrench. The tub can then be removed. The procedure that follows is for replacing the more modern kind of tub that resides in an alcove and is abutted on three sides by walls:

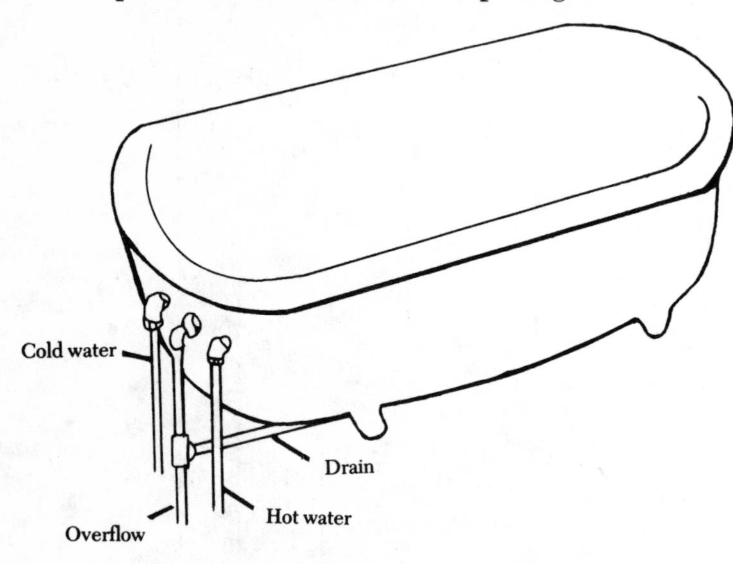

Old-fashioned tubs have the advantage of exposed plumbing lines that are easy to reach.

1. Close the hot and cold water supply valves and open the tub faucets to drain all water from the pipes. The supply lines are not connected to the tub itself, but the drain is, and in order to reach it you may have to break through the back of the wall that houses it. If the tub is on the ground floor, check the basement ceiling beneath it to see if you can reach the drain connection. Also, examine the wall behind the tub for an access panel. If you find no way of getting at the drain, try to disengage the tub at its strainer and overflow lever before you knock any holes in the plaster.
2. Unscrew the escutcheon surrounding the stopper control handle and remove the stopper mechanism.
3. Unscrew the strainer basin. The basin may have become so corroded that it may seem as if it is not even threaded. But it is. If there are small notches in the insides of the basin and you have a spud wrench, insert it and try rotating the basin counterclockwise. If that fails, jam a long screwdriver between the bars of the drain and try to pry or hammer the strainer loose.
4. If you can free the strainer basin and the overflow control handle, you have disconnected the tub from the drain line and can proceed to remove it from its niche. If you cannot free them, you have to get at the back of the tub, which may mean breaking through the wall and then freeing the drain line at whatever fittings you can loosen.
5. Bathtubs stand on the floor with their wallside rims supported by a 1″ × 4″ board nailed to the wall studs. They are also caulked around the top of their rims and their base. The caulking may have to be

chipped loose before the tub can be pried out of its niche. Beware that tubs can weigh anywhere from 50 to 500 lbs., and if nothing else they are awkward objects to handle—even when two persons are working with them. Pull the tub away from the walls and disassemble its waste and overflow fittings from the drain line.

6. The old supports may fit under the replacement tub. More likely, they must be pried off the wall studs and either repositioned or replaced. Stand the new tub on the floor and measure the distance from the floor to the underside of the back and end rims. Remember that large metal castings such as bathtubs are not necessarily perfect in their dimensions. If the three measurements are not identical, use the highest one to mark the walls. Strike a level line along all three walls at the point of your highest measurement. In theory, you want to position the tub so that it is absolutely level in all directions; but if it must be out of level, make sure it pitches toward its drain end.
7. Nail 1″ × 4″ support boards to the studs along the three level lines you have marked on the walls.
8. Slide the tub into its niche and rest the back and end rims on the support boards.
9. The base of the tub should be resting squarely on the floor. If it is not fully supported, place shims under the base to level the tub as perfectly as you can get it.
10. When you have leveled the tub, mark the center of both the drain and overflow holes on the wall.
11. Remove the tub from its niche and nail the shims in place.
12. If you have enough space to work in, you can put the tub back in its recess and then assemble the drain fittings and make your connections to the drain line. If not, assemble the waste and overflow fittings to the drain line in the wall so that the center of the control handle and the center of the drainpipe are over your markings on the wall and floor. To accomplish this, you may have to move the existing drainpipe

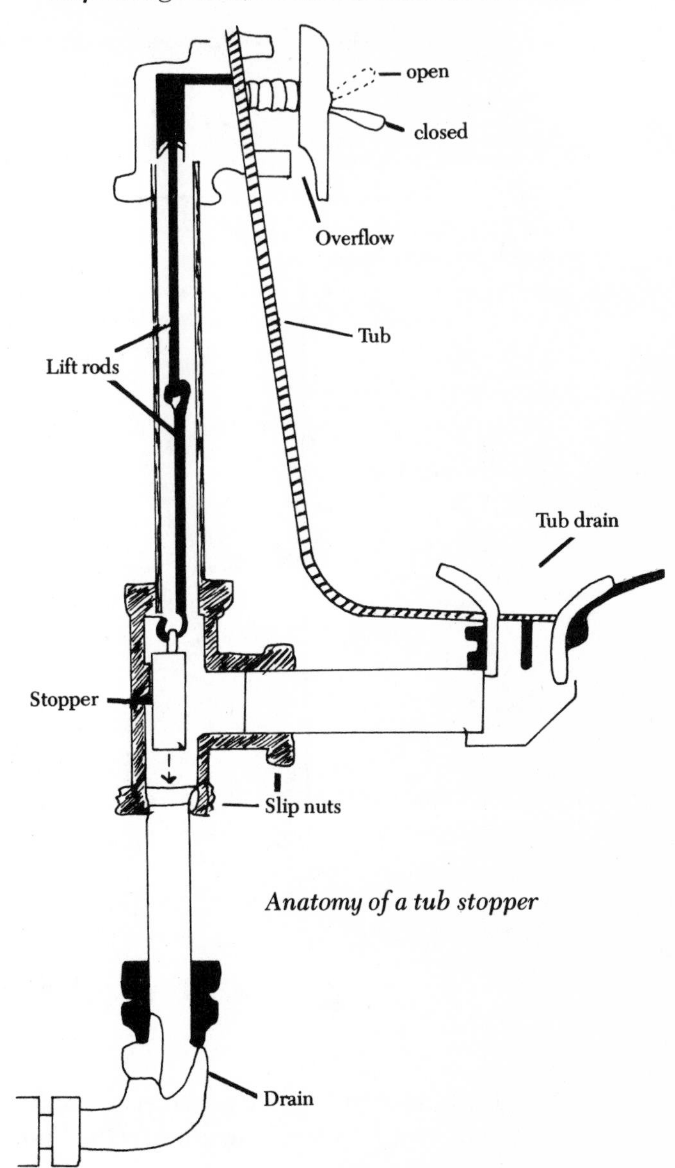

Anatomy of a tub stopper

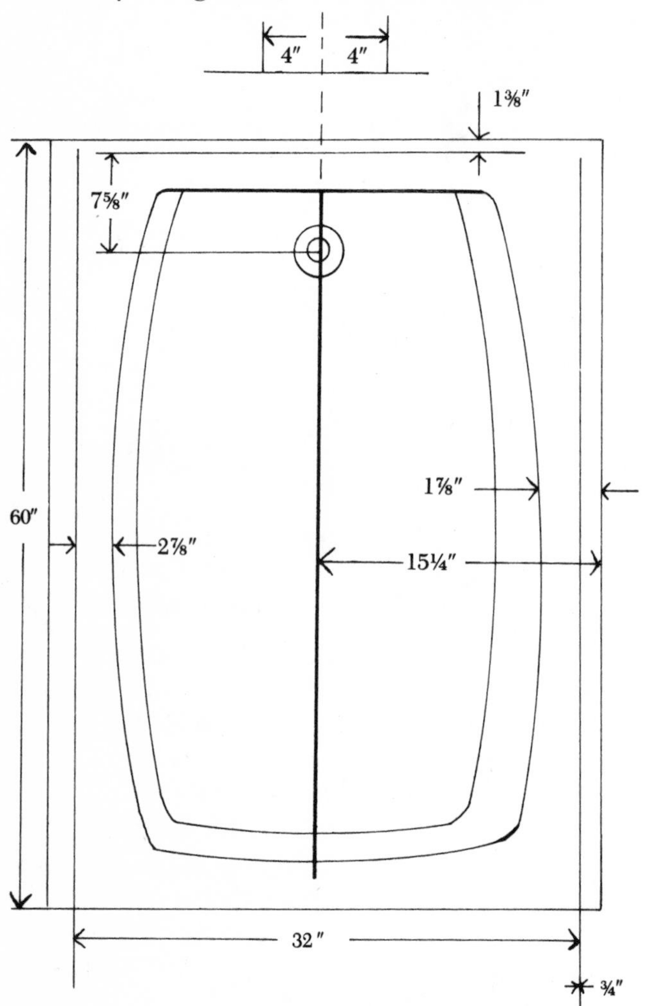

slightly, or even replace portions of it until the tub waste fittings align perfectly with your markings.

13. When the waste and overflow fittings are in place, put the tub into position. If your measurements have been accurate, you should be able to screw the drain basket into the drainpipe and the control handle escutcheon into its hole. The drain basket should have its threads wrapped in plastic pipe-sealing tape or plumber's putty before it is assembled; the control handle need not have a watertight seal.
14. Turn on the water supply and run water in the tub. Examine all of your connections for leakage and tighten any fittings that show signs of water seepage.
15. Apply a bead of caulking around the rim of the tub and its base.

Dimensions for a typical tub installation

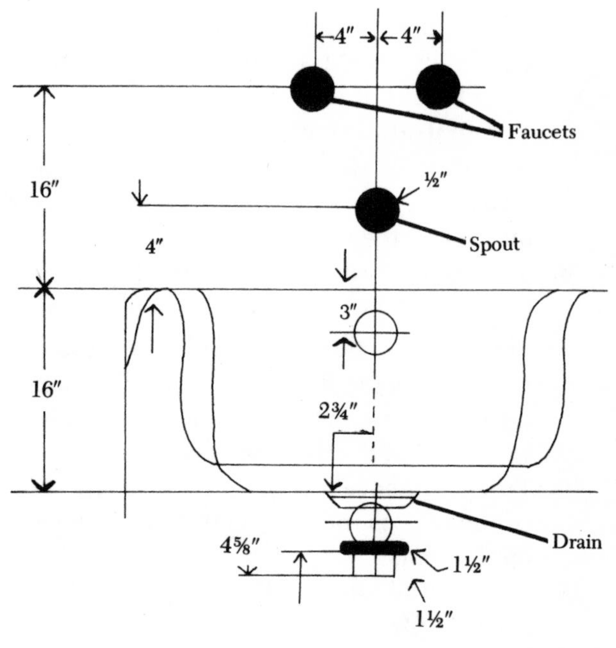

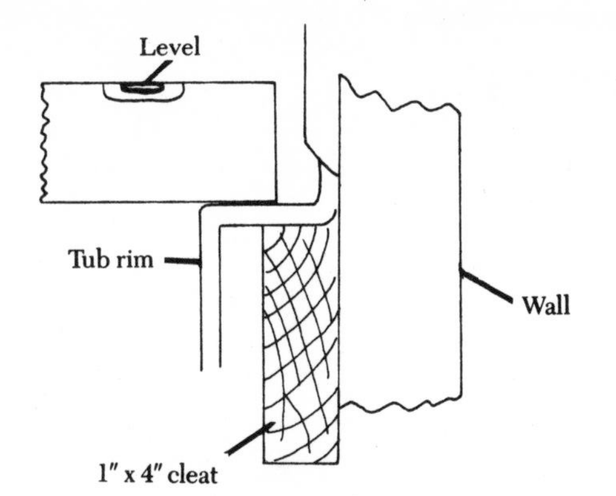

The tub must be level in two directions. You can place a board on edge across the length of the tub and rest your level on top of it to get an accurate reading. But be sure the board is straight.

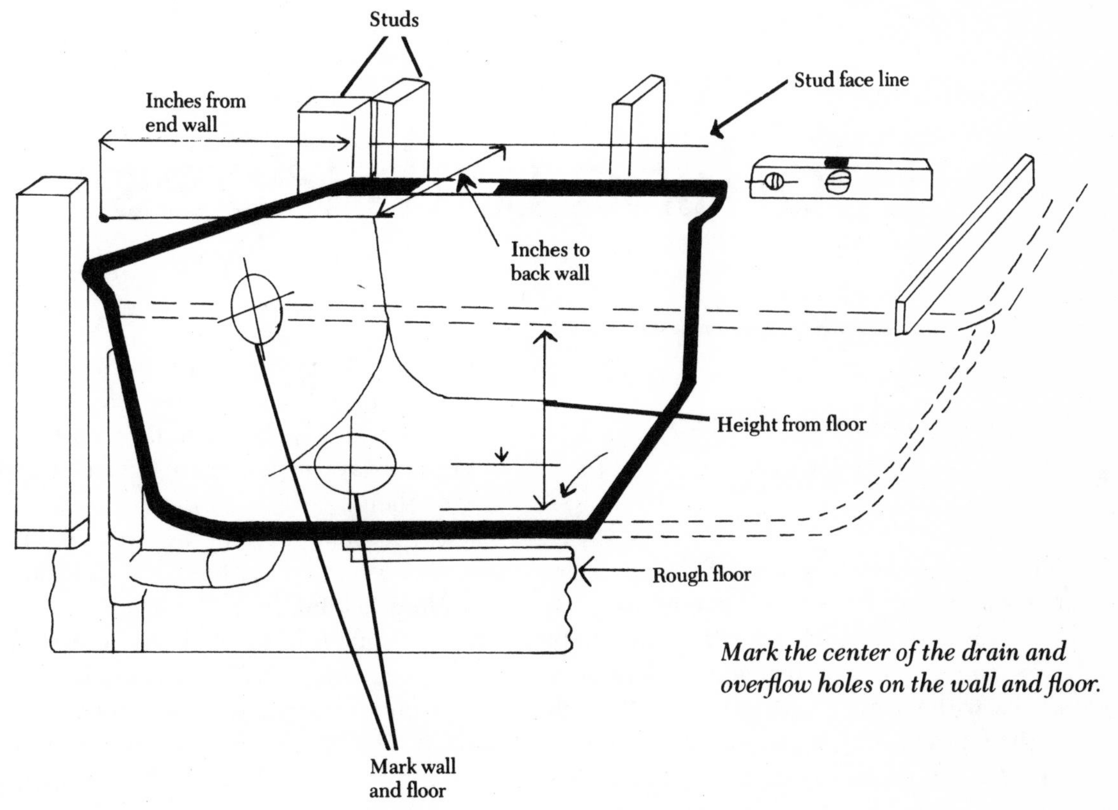

Mark the center of the drain and overflow holes on the wall and floor.

CHAPTER 5

Working with Pipes

Understanding Your Plumbing System

With the possible exceptions of changing the drain hookup under a sink or extending your water supply system by a few feet to incorporate a shower stall, chances are that you will lead your entire life without ever having to wrestle with any of the pipes in your home plumbing system. Nevertheless, it doesn't hurt to understand the purposes of all those pipes that hang from your basement ceiling, disappear into your walls, and then mysteriously reappear in your bathrooms and kitchen.

In a nutshell, the plumbing system in every home is made up of two separate systems. One of those is the network of water supply lines that bring water into your house under a constant pressure of 40 to 100 lbs. per square inch (p.s.i.), and carry it to every faucet in the building. The water supply lines are always filled with water, and the water is always under pressure, which is why it always gushes out of the faucets when you open them.

The second system in every home is the drain-waste-vent system (DWV), which consists of pipes leading from the lowest point of every fixture (sink, tub, toilet) in the house to a vertical soil pipe, which in turn leads to a horizontal pipe in or under your basement known as the **house drain**. The house drain carries all of the water in the fixtures out of the house to a municipal sewer or private waste-disposal field. House-drain pipes are rarely ever full of anything except air, which enters through the soil stack and perhaps several other vents that extend up through your house and out the roof. The venting portion of the system exerts atmospheric pressure on the water running through the drain lines to force it to continue traveling downhill and out of the house.

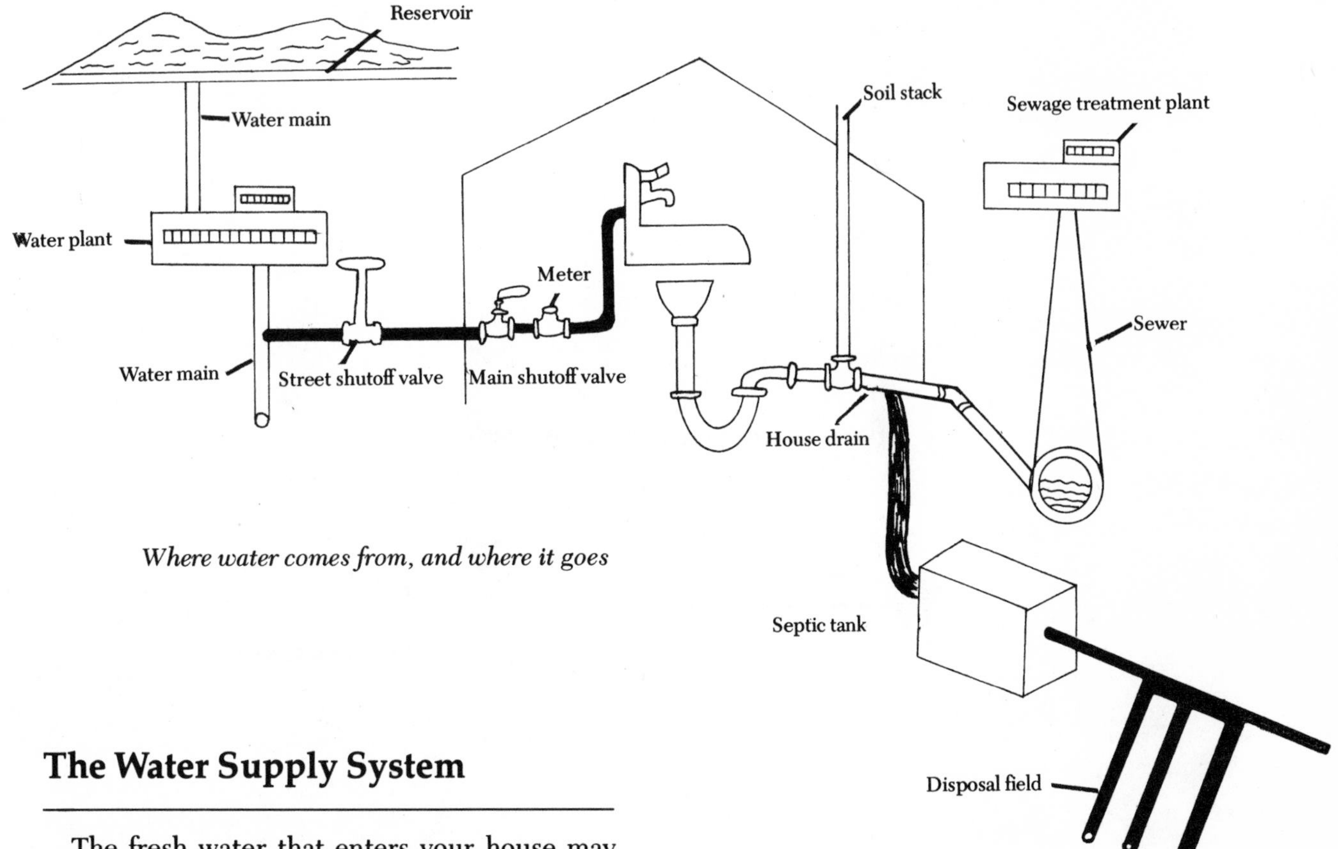

Where water comes from, and where it goes

The Water Supply System

The fresh water that enters your house may come from a municipal water system, a stream, a well, a reservoir, or a storage tank. It arrives at your house via a 1″-wide entrance pipe that may be made of lead, copper, or brass. Should anything happen to the entrance pipe, it is usually the responsibility of the municipality to repair it; however, as the homeowner, you may have to pay part or even all of the repair costs.

In your basement, the entrance pipe ends at a gate or globe valve (the **main shutoff valve**), which has the sole function of closing off all of the water coming into your home. Just past the main shutoff valve you will discover a water meter (unless it is positioned outside your house), and there may be a pressure-reducing valve that regulates the pressure in your water supply system to about 60 lbs. p.s.i. If the water pressure is inadequate, you can usually increase the pressure by resetting the valve. If the valve is not functioning, it can be replaced.

The water main that leaves your water meter is, in most instances, a ¾″-wide pipe, although it can be ½″ wide if your home is small. The meter marks the beginning of your cold water main; if you follow it along your basement ceiling you will find that it goes directly toward your hot water heater, at which point there is a branch line leading to one side of the heater. The other side of the heater has a ¾″- or ½″-wide hot water main leading away from it, which goes up to the basement ceiling and then runs parallel with the cold water main. From the hot water heater on, the two water mains extend as directly as possible to each of the sinks, tubs, and toilets connected to the system.

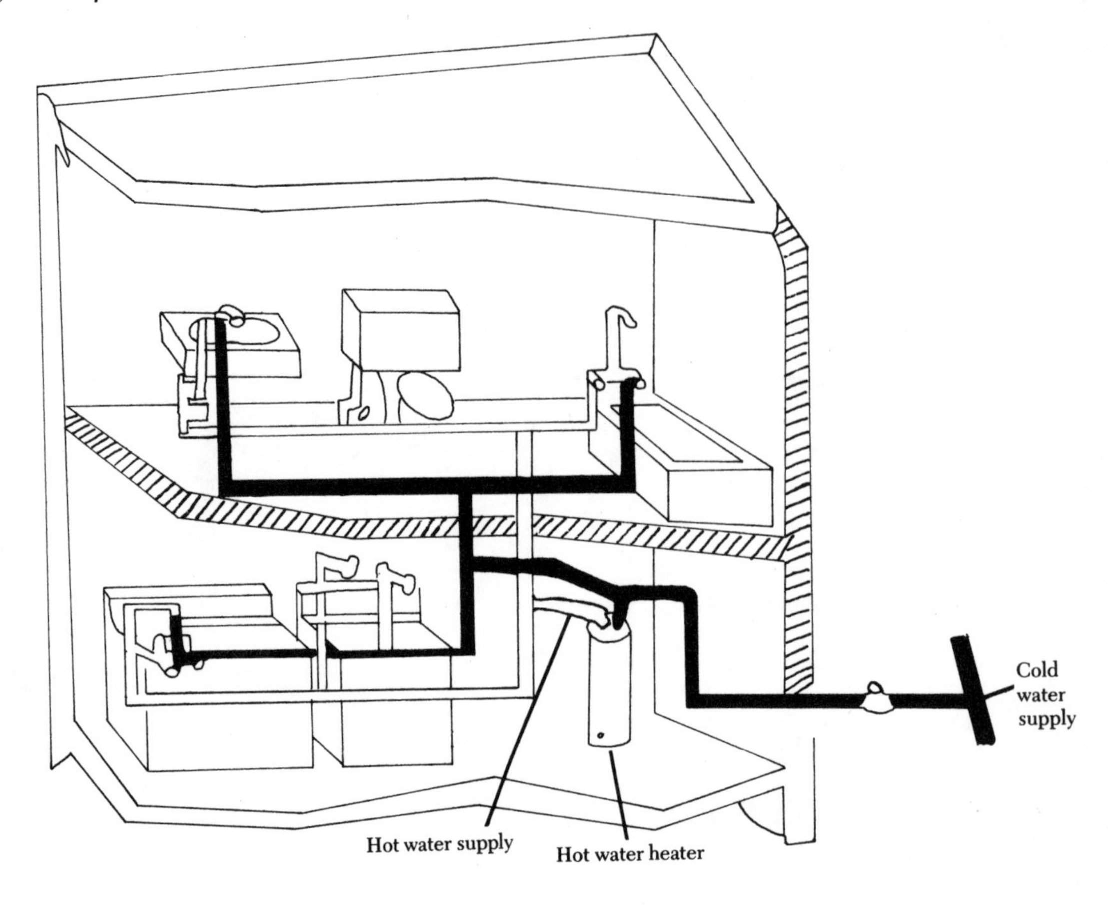

The water supply system

As the supply mains approach a fixture, a separate ½″-wide pipe branches away from the main and extends through the wall near the fixture. An air chamber is attached to this pipe just before it leaves the wall. Every faucet should have an attending air chamber, which is a 12″ to 18″ piece of capped pipe that extends vertically up the inside of the wall. Every time you open and then close a faucet, you are stopping a rush of water that is about 60 lbs. p.s.i., which is like applying the brakes on a moving car; unless there is some way to cushion the shock, the water will rattle the pipes all over the house as it comes to an abrupt halt. That cushion is the air

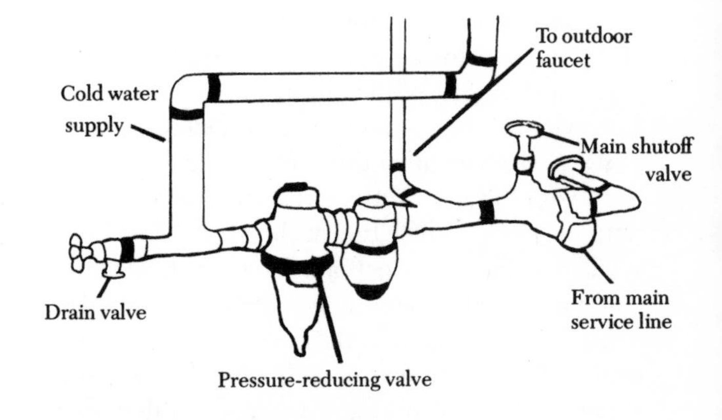

There may be a pressure-reducing valve positioned near the main shutoff valve.

chamber. When you shut off a faucet, water backs up into the air chamber and is cushioned by air. If the pipes in your home rattle and clank every time a particular faucet is closed, you probably do not have an air chamber at the faucet. The noise is known as **water hammer**, and it can be dangerous enough to blow your entire plumbing system apart.

Air chambers can be added to your system outside the walls merely by attaching them to the feeder line behind the fixture shutoff valve. An air chamber can be a simple, straight piece of pipe with its end capped off, or one of the many more intricate designs available at home hardware stores or plumbing supply outlets.

Once the branch line has diverted from the supply main and emerged through the wall, it ends at a fixture shutoff valve. The top of this valve has a socket that accepts a length of flexible ⅜″-wide copper or plastic tubing, which leads to the tailpipe of the faucet or, in the case of a toilet,

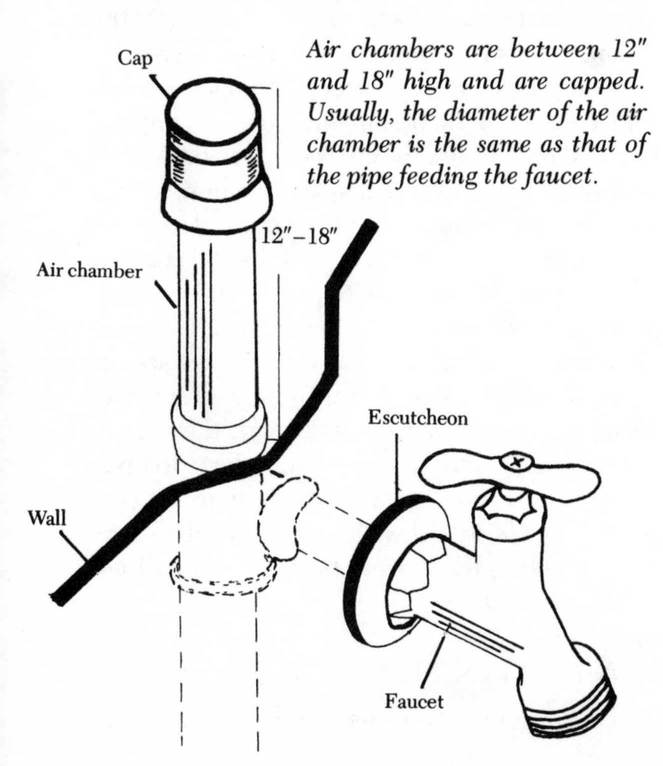

Air chambers are between 12″ and 18″ high and are capped. Usually, the diameter of the air chamber is the same as that of the pipe feeding the faucet.

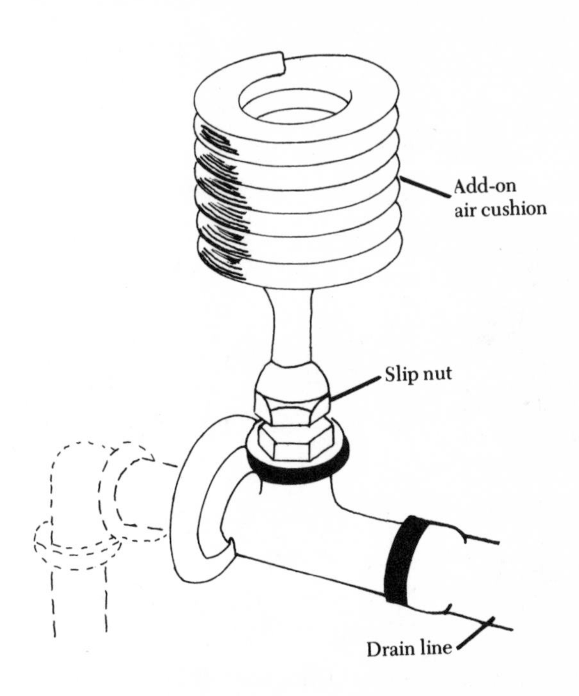

Add-on air chambers are sold in a variety of designs, all of which work well. You can insert them behind some faucets or, more often, directly behind the faucet shutoff valve.

to the flush valve. Not all faucets have separate shutoff valves; ideally, they should. Any time you have occasion to work on a fixture that has no shutoff valve of its own, you should add one so that in the future you can shut off the water to that fixture without closing off service to the rest of the house. The addition is a simple procedure.

There is one cardinal rule concerning a water supply system: *Every connection, every fitting, must be watertight and capable of withstanding the constant pressure of water racing through the pipes.* This means that the connections must always be soldered if the pipe is copper, or solvent-welded if it is plastic. If the pipe is galvanized steel or brass, the threads must be caulked with plastic pipe-sealing tape or plumber's putty.

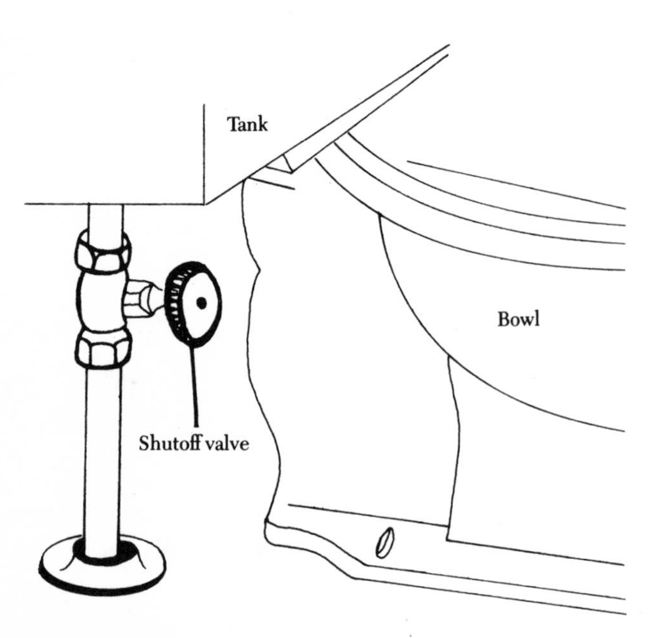

Toilets nearly always have shutoff valves; many faucets do not. The valve can easily be added to the supply line as soon as it comes out of the wall or floor.

The Drain-Waste-Vent System

The drain-waste-vent system in your home is absolutely separate from the water supply lines. In fact, a great many precautions are taken in the design of the DWV to prevent any of the waste that flows through the system from backing up into the supply lines and contaminating your fresh water.

The purpose of the DWV is to carry waste away from the fixtures and out of your home. It must be designed so that neither vermin nor sewer gases can travel through the empty pipes and into your house. Thus, there are traps connected to the drain line of every sink, basin, shower, and bathtub in your home. Toilets, because they always have water in their bowls, are natural traps.

A drain trap is either bowl-shaped (as under a shower stall) or P- or S-shaped. Drum traps, which were popular 40 years ago, are still found in many older homes, but these have proven to be breeding grounds for vermin and are now prohibited as health hazards in most residential plumbing systems. If you have any drum traps in your existing system, replace them the next time they clog.

A trap is supposed to hold water in the bottom of its lowest curve. It is connected at one end to a tailpipe under a fixture, and to a waste pipe at the other end. When repairing an old drain, you may encounter a problem with pipe size: the new, standardized size is 1½″ in diameter, while the pipes in many older systems are 1¼″. If you need to make connections with new pipe, you will need a reducer coupling that is 1¼″ at one end and 1½″ at the other.

The waste pipe slopes away from the fixture and connects to a vertical 3″ or 4″ vertical pipe known as the **soil stack.** The soil stack is traditionally cast iron, although many modern constructions incorporate plastic. The stack extends from the roof to the basement, where it connects to the 3″ or 4″ house drain. The house drain runs horizontally from the soil stack to wherever the outside sewer connects to the house. The drain has a cleanout plug at the base of the soil stack, at all turning points of more than 45°, and at a final point just before it exits the house. Cleanout plugs are threaded caps that screw into a special fitting in the pipe run. Because they often go for decades without ever being opened, they sometimes "freeze" shut and resist all efforts to open them. Your recourse when this happens is to cut out the entire fitting with a cold chisel and replace it with a lead or plastic plug that will not weld itself shut.

VENTING IN THE DWV

The purpose of venting a drain system is to

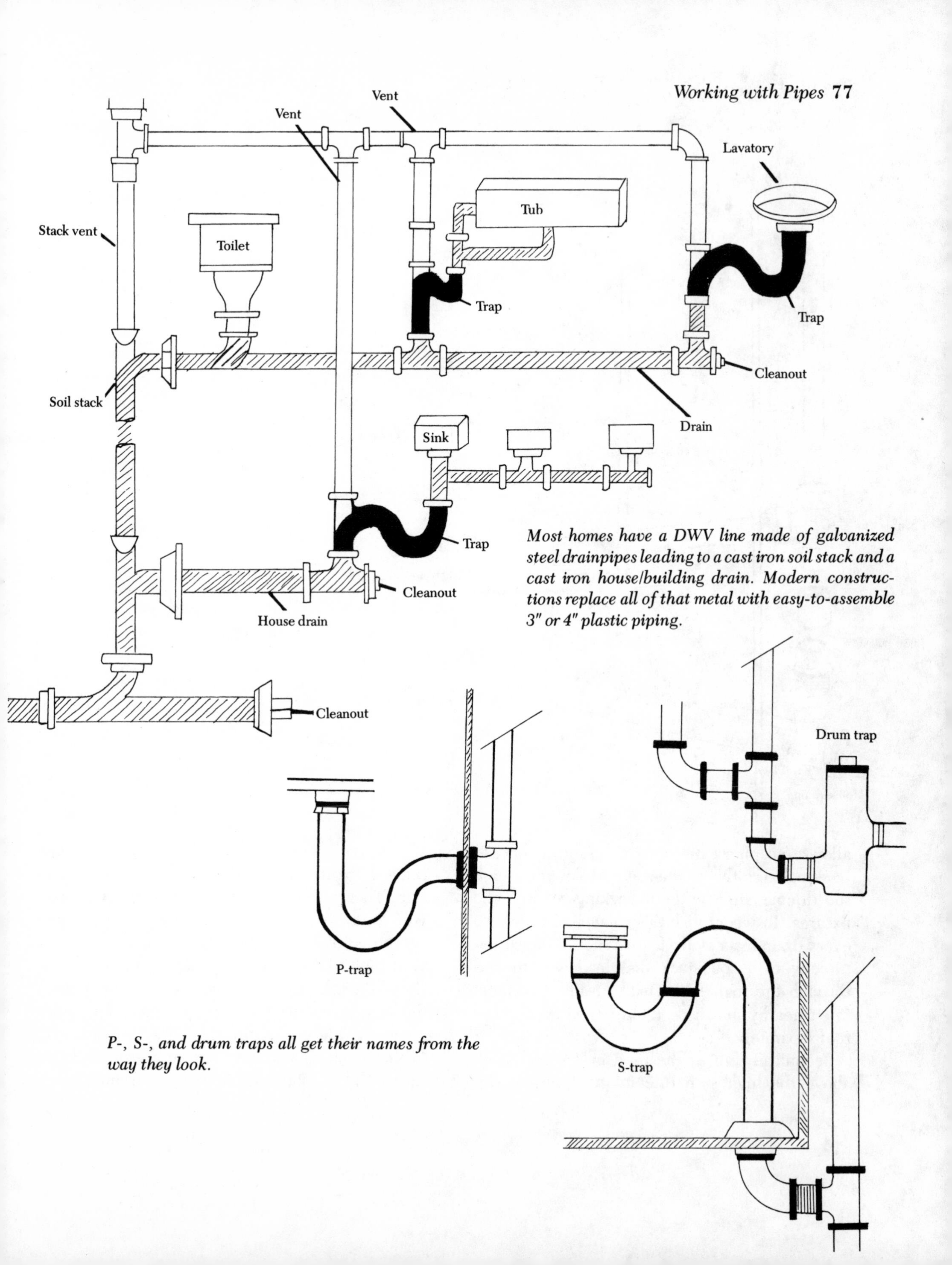

Most homes have a DWV line made of galvanized steel drainpipes leading to a cast iron soil stack and a cast iron house/building drain. Modern constructions replace all of that metal with easy-to-assemble 3" or 4" plastic piping.

P-, S-, and drum traps all get their names from the way they look.

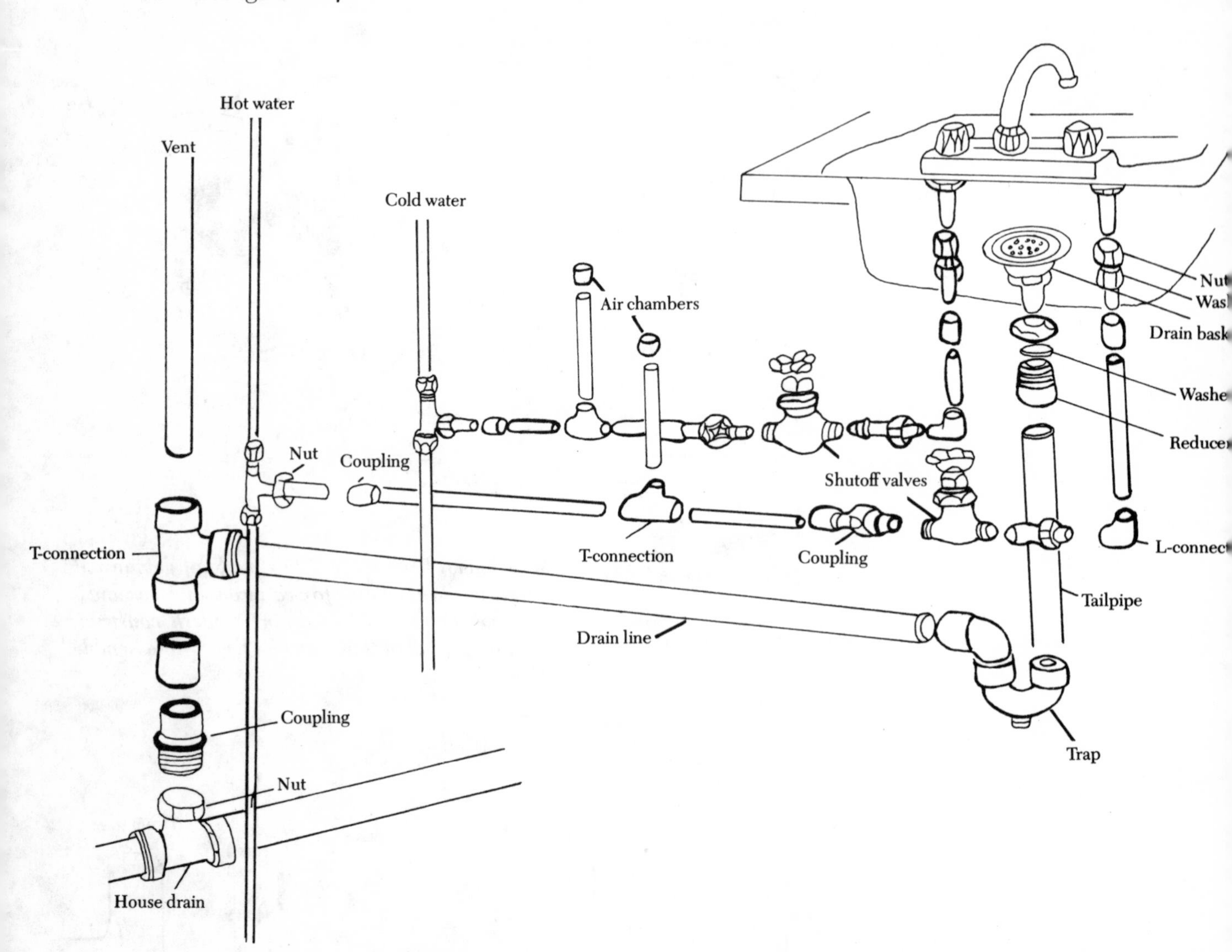

allow atmospheric pressure in the supply lines to be equaled so that waste can flow down through the drain system without backing up into the fixtures. To accomplish this balance of pressure, every drain from every fixture should be vented directly to a soil stack that leads up to and through the roof. If the fixture cannot connect into a nearby soil stack, it must be vented to the roof separately.

Technically, all of the pipe in the soil stack above the highest fixture in the house is designated as a vent, while all of the pipe below the highest fixture is considered part of the waste line. There are two kinds of vents: **revents** and **wet vents**. Whenever a plumbing fixture cannot be vented directly to the outside, it may have a vent pipe that begins at its trap and extends to the soil stack. But to avoid cross-connections, the vent pipe must join the soil stack at a point *above* the highest fixture that directly connects to that stack. If you are venting a ground-floor sink to the soil stack, for instance, you may have

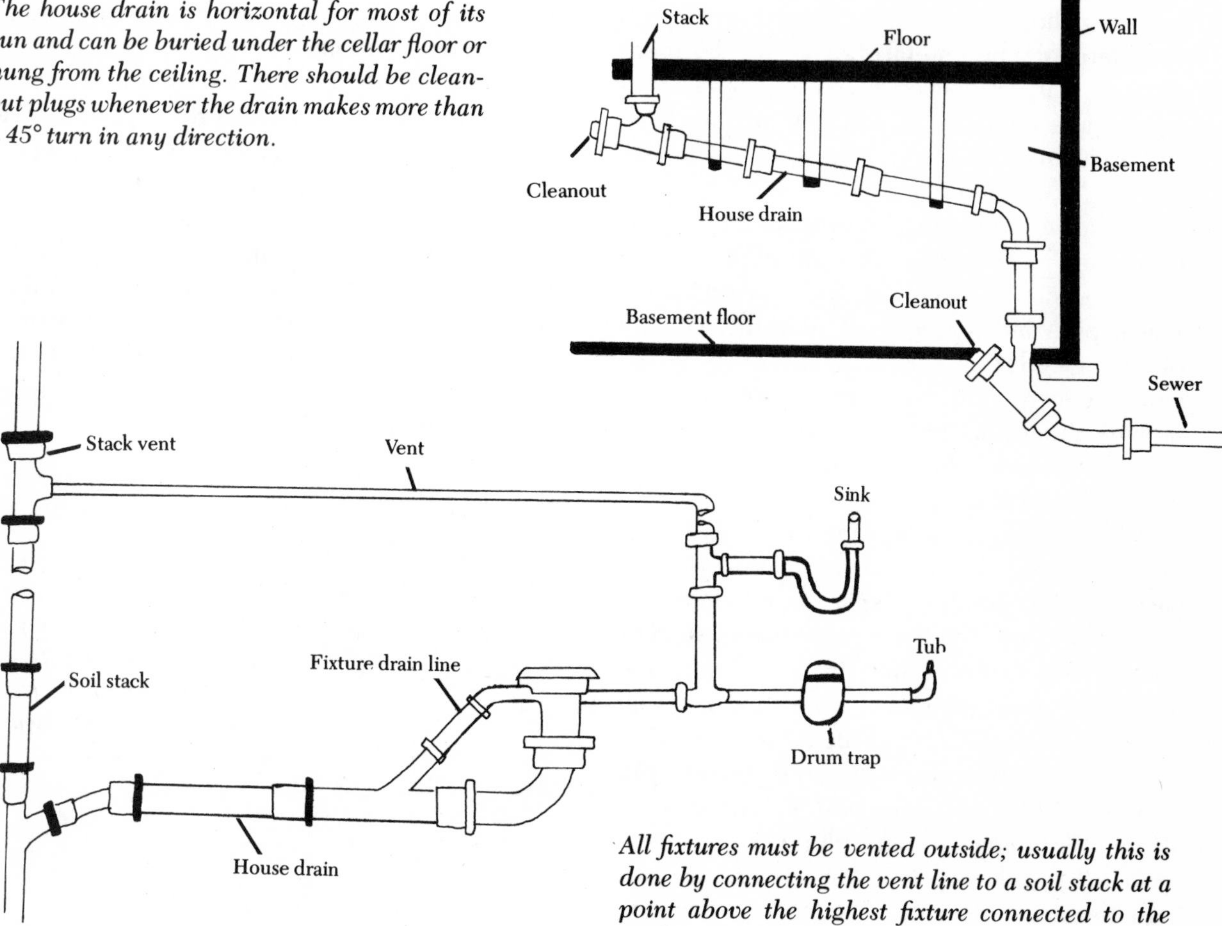

The house drain is horizontal for most of its run and can be buried under the cellar floor or hung from the ceiling. There should be clean-out plugs whenever the drain makes more than a 45° turn in any direction.

All fixtures must be vented outside; usually this is done by connecting the vent line to a soil stack at a point above the highest fixture connected to the stack.

to run the vent pipe up to a point above the second-floor bathroom sink before it can attach to the soil stack. This kind of assembly is called a **revent**.

It is also possible to use the fixture's waste pipe as a vent, in which case the pipe is known as a **wet vent**. The national plumbing code has established specific diameters for wet vents that depend on the length of the pipe before it connects to the stack. Thus, the longer the pipe run, the larger its diameter must be, so that it can never completely fill with water and cause a suction that will put inordinate pressure on the seals around the trap.

Pipes and How to Work with Them

There are four kinds of pipe widely used in

residential plumbing systems: galvanized steel, brass, copper, and cast iron. Galvanized steel is probably the most common. Brass, while it is considered the best metal of all, is now so expensive that it is rarely used anymore. Copper piping is quite common in water supply systems, and cast iron is used exclusively in house/building drains and soil stacks.

The newest and most versatile kind of pipe is plastic: it is less expensive than any of the metals, lighter, easier to assemble, and more durable. Plastic pipe and tubing meet all of the national plumbing codes and are fully approved by the federal government; there are, however, a few municipalities that still prohibit their use. Be sure to check your local plumbing codes before you decide on plastic.

Plastic pipe enables anyone to do his or her own plumbing repairs and installations—even without prior plumbing experience. While a lot of practice is needed to properly thread steel or brass pipe or solder copper fittings, plastic pipe and tubing requires practically no plumbing skill at all. Plastic is the wave of the future in plumbing that you can work with now—quickly and inexpensively. And it's the best reason for doing all your plumbing without a plumber.

So that you can repair and improve your existing plumbing system, this section will show you how to work with the various metal pipes as well as with plastic ones. But before you begin, there are some general principles and procedures you should know.

MEASURING PIPE

Pipes are always referred to by the measurement of their inside diameter. However, a pipe that is nominally ½", ¾", or 2" in diameter might actually be slightly larger or smaller than that. When you are determining the size of any pipe, measure its inside diameter and don't be surprised if it is not a standard measurement. If you are buying pipe that must connect to existing lines in your plumbing system, whenever possible take a fitting or piece of the pipe with you when you go shopping, just to make sure you get the diameter you need.

To measure the correct diameter of the inside of a pipe, use a rigid ruler. Press the end of the ruler against the inside of the pipe and swing it in an arc across the hole, and read the distances against the opposite edge of the hole. Your readings will increase, and then decrease. Use the longest measurement you get as the nominal diameter of the pipe. The most common diameters are ¼"; ⅜"; ½"; ¾"; 1"; 1¼"; 1½"; 2"; 3"; and 4", but your measurement could vary by as much as a quarter of an inch.

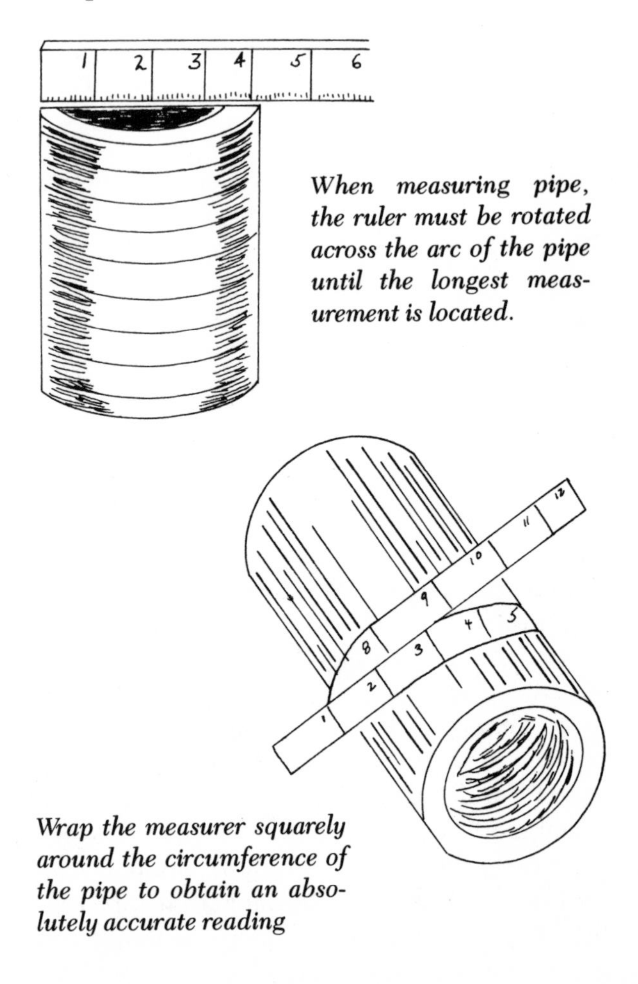

When measuring pipe, the ruler must be rotated across the arc of the pipe until the longest measurement is located.

Wrap the measurer squarely around the circumference of the pipe to obtain an absolutely accurate reading

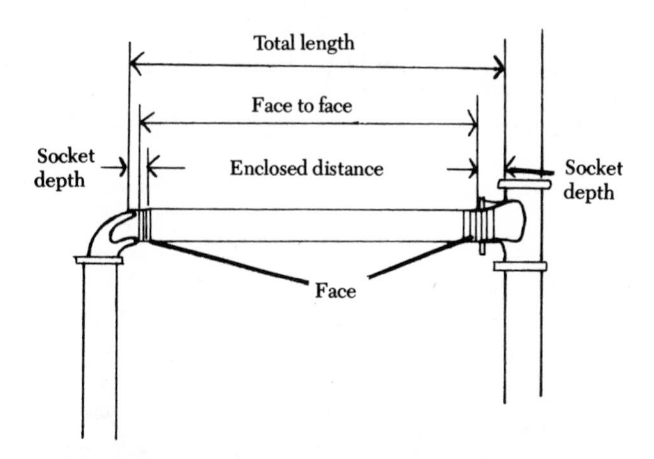

The make-up distance is the amount of pipe that will be inserted in the fittings at each of its ends.

Sometimes you cannot get at the end of a pipe to measure its nominal diameter, so you will have to measure its circumference. Any plumbing supply outlet has a chart that will convert the circumference to the proper nominal diameter. When measuring the circumference of a pipe, use a flexible rule, such as a steel tape, that you can wrap around the pipe. Wrap the rule squarely around the pipe or you will arrive at too long a measurement. (A steel tape rule is also useful in measuring pipe length for cutting, since it has a tab attached to one end. Hook the tab over the end of the pipe and draw the tape tautly along the pipe to the point where you wish to cut it.)

Before you cut any pipe, you must not only measure the distance you want it to be, but add to that length a **make-up distance**. The make-up distance is the amount of pipe that will reside inside the fittings attached to each of its ends. In order to know exactly how much of the pipe will go into the fitting, you should measure both fitting sockets. The sockets can vary as much as nominal pipe diameters, so never assume what their depth will be. Always measure them. Even with all of this careful measuring, the safest procedure is to arrange the pipes to be cut with the fittings that will be attached to them, and then measure everything again before you start the cutting. Piping can be quite expensive, and you have very little margin for error.

FITTINGS

Fittings are odd-shaped pieces of pipe used to connect longer lengths of pipe together. Each of the different kinds of pipe used in conventional plumbing systems has a full range of fittings available that will allow you to do just about anything you want, from curving around a corner or branching off in different directions to changing the size of your pipe in the middle of the run. While the shapes of the fittings are generally the same, there are two specific types: **standard** and **flush-walled**. Standard fittings are used in water supply lines and the vent portion of your drain system. They are designed so that a ridge, or shoulder, is created by the end of the pipe when it is inserted in the fitting. This shoulder does not disturb the flow of clean water through the pipes in any way, but it does trap debris. The flush-walled fittings are designed to provide a smooth inside joint so that waste flowing through the drain lines has nothing to obstruct it.

Fittings are so numerous in design and shape that rather than trying to memorize them, simply go to your plumbing supplier and explain the job that you are trying to do. The salesperson probably has a chart of all the fittings available, and should be able to help you make the proper selection.

Branch fittings are called **T**'s or **Y**'s because of the way they look. They have three sockets used to join two pipes that are going in the same direction and a third pipe that intersects at either a 90° or a 45° angle. The third pipe can be a different diameter from the other two. When buying a branch fitting, state the "run" or

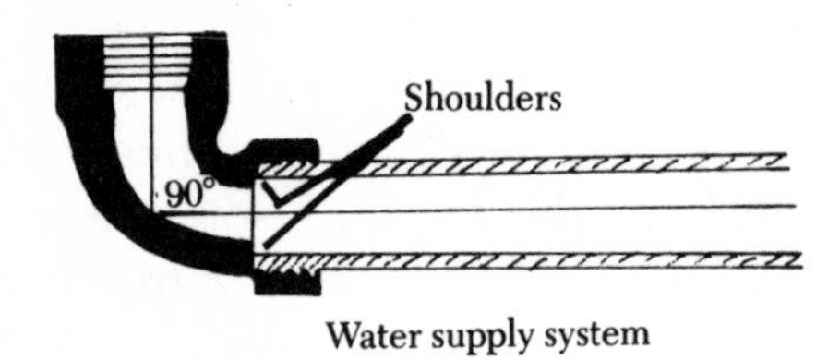

Standard fittings have a small ridge inside their connections.

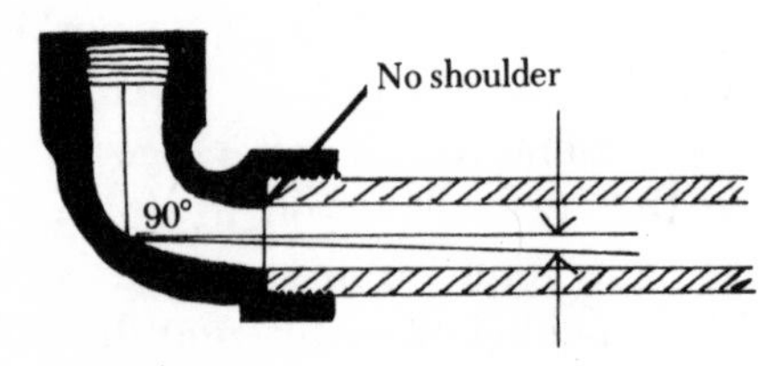

Flush-walled fittings have a smooth inner surface at their connections.

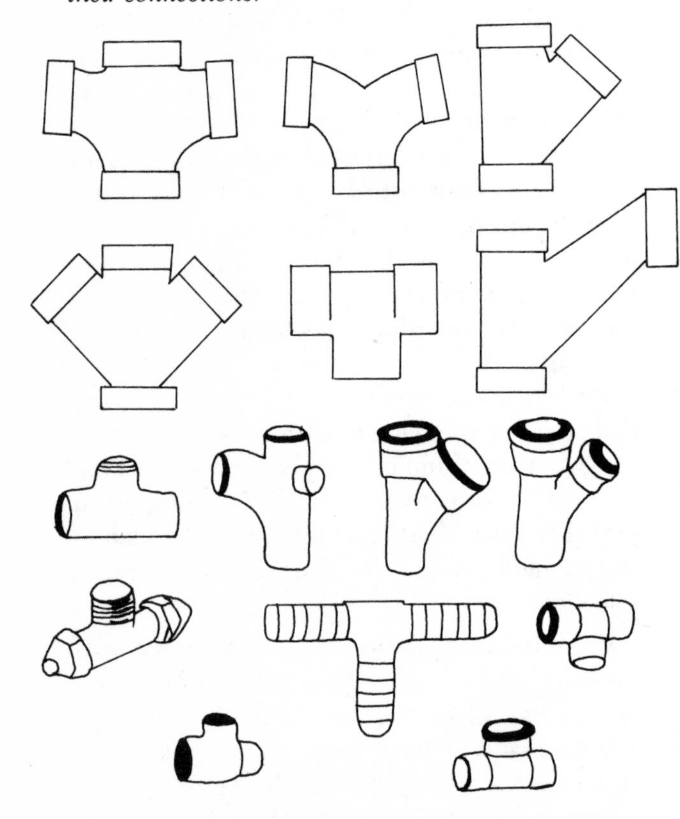

Branch fittings (T's and Y's)

"through" size first,and then the branch size. Thus, for a T that allows a ½″ pipe to intersect a ¾″ pipe run, you would ask for a ¾″ × ¾″ × ½″ T.

Elbows join pipe at either a 45° or 90° angle.

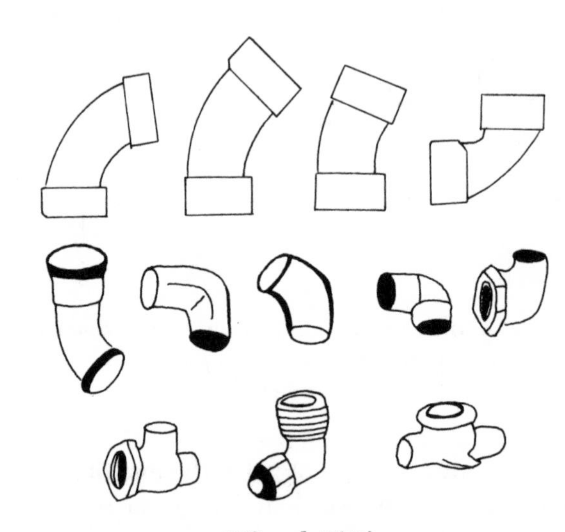

90° and 45° L's

Couplings connect two pieces of pipe in a straight line and are used when a length of pipe is not long enough, or when the pipe may need to be disassembled at a later time.

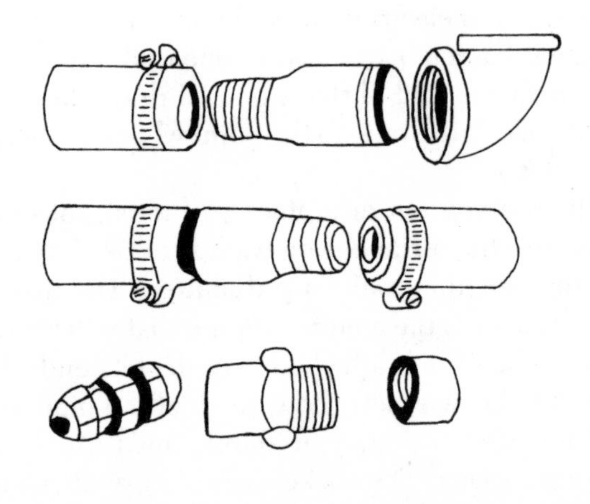

Couplings

Slip couplings are used in DWV systems to add fittings to an existing pipe, and have no inner shoulders.

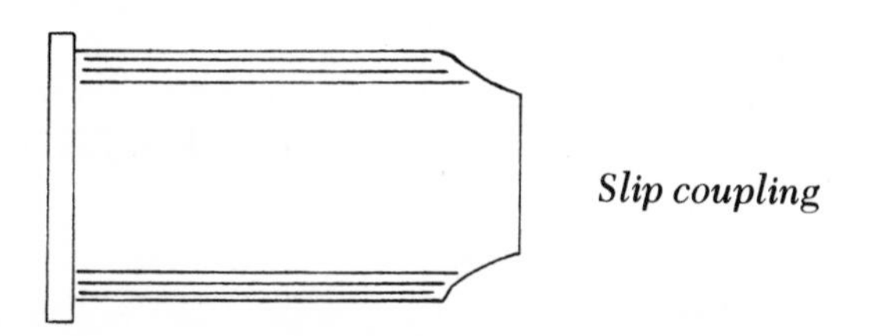

Slip coupling

Wing (Drop-ear) elbows are designed to support shower arms or faucets; their ears are fastened to a framing member.

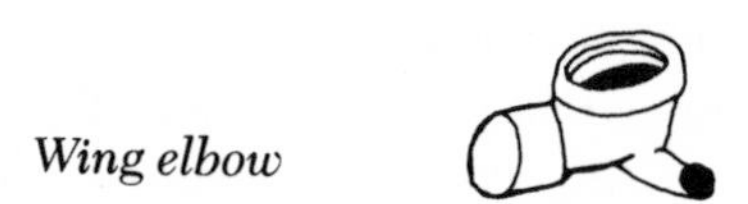

Wing elbow

Reducer couplings accept a pipe of one diameter in one end, and a different diameter pipe at the other. They can either be T- or Y-shaped.

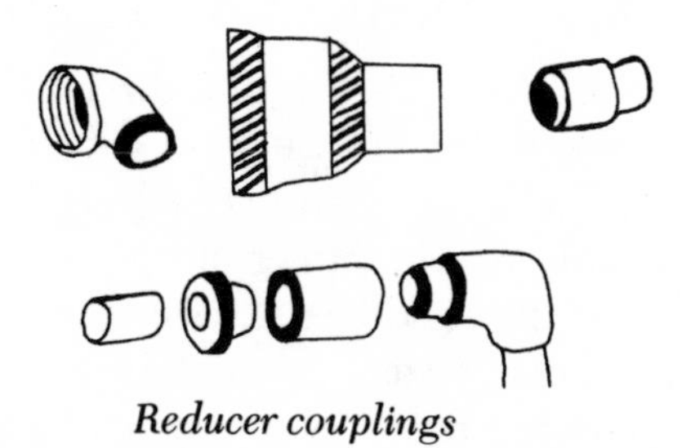

Reducer couplings

Bushings allow you to connect different-sized pipes and fittings by nesting into the larger fitting socket at one end and accepting the smaller pipe at the other.

Bushing

Crosses permit four pipes to come together from opposite directions.

Cross T

Dielectric couplings connect water supply pipes of different metals so that there will be no undesirable electro-chemical reaction, such as corrosion.

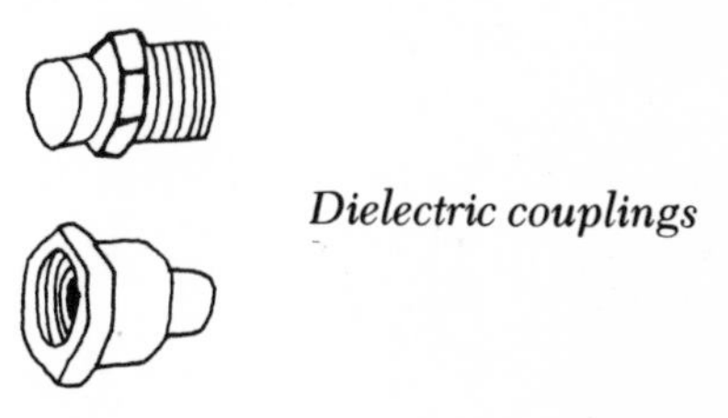

Dielectric couplings

Unions connect any pipe that you expect to take apart at some future time.

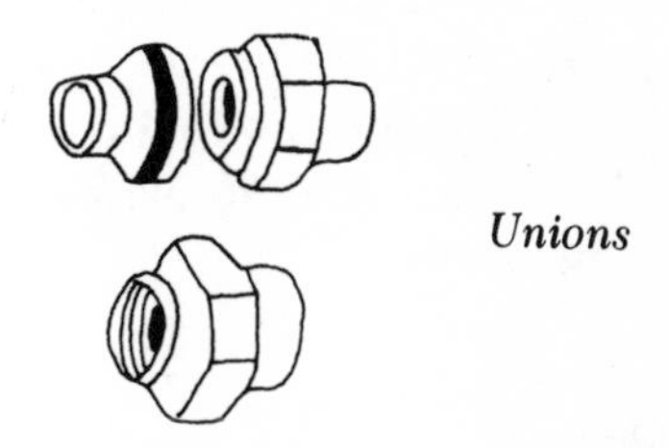

Unions

Plugs are placed in the socket of a fitting to seal it.

Plug

Caps are placed over the end of a pipe to seal it.

Caps

Adapters are used to join two pipes or fittings of different types.

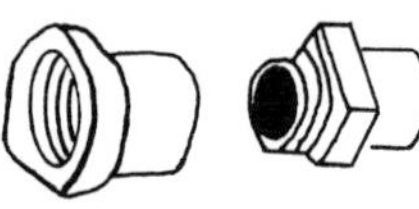

Adapters

Nipples come in various diameters and lengths ranging from about 1″ to 12″.

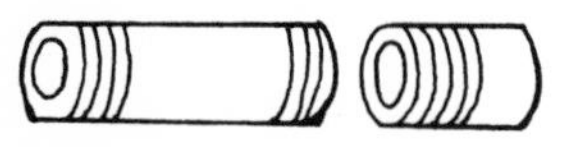

Nipples

Cleanouts are Y-fittings with removable plugs that give easy, quick access into a DWV pipe for removal of a stoppage.

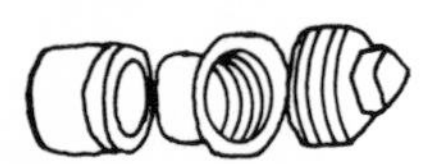

Cleanouts

Street fittings —"Street" means pipe-sized. Street fittings are used both in water supply and DWV lines, and have one end that is pipe-sized to nest directly in a fitting socket without a short nipple placed between.

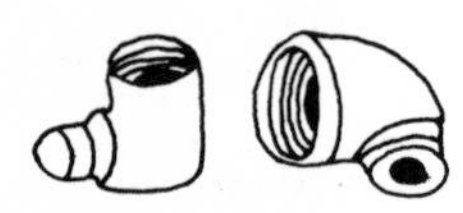

Street L's

CONNECTING FITTINGS TO THREADED PIPE

Threaded connections, whether you are making them with plastic or metal, involve some elements that you should always keep in mind. If you examine the threaded portion of a pipe closely, you will notice that it is actually conical in shape, growing a millimeter or so with each thread. As a result, the threads create a seal that is fairly watertight. But in order to ensure a connection that is absolutely watertight, you must first coat the threads with plumber's putty, plastic pipe-sealing tape, or "pipe dope" (a kind of liquid putty that is applied with a brush). The pipe can then be screwed into the fitting socket until it is tight. But be careful not to overtighten: too much winding might strip the threads or even split the fitting. As a rule of thumb, you have tightened a fitting enough when you can still see three threads of the pipe. Remember that if you thread a fitting in place and it leaks at a joint, you can probably stop the leak by tightening the connection another half turn. The best approach is to hand-tighten the connection and then give it a half turn more with your pipe wrench, and then do no more tightening unless you discover that the connection leaks.

When using two pipe wrenches, always apply them in opposite directions. One wrench should be around the pipe to steady the work. The second wrench goes around the fitting with the open side of the jaws facing the direction in which you will be turning the wrench. The fitting is rotated clockwise to tighten and counterclockwise to loosen.

REPAIRS TO THREADED PIPE

Anytime there is a leak at a threaded fitting, tighten the fitting or the pipe a half turn by rotating clockwise (secure the pipe with one wrench and turn the fitting slowly with another wrench). If the leak does not stop, the fitting is probably damaged and should be replaced.

If tightening does not work, or if there is a small crack in the pipe or the fitting, you may be able to seal the leak with epoxy cement. However, the epoxy will make disconnecting the joint at a later time somewhat difficult.

Galvanized Steel and Brass Pipe

Galvanized steel is inexpensive and can withstand tremendous pressure, but it also corrodes easily and will become clogged with the scale left by "hard" water (water that contains large deposits of minerals). Brass is the best type of metal piping you could possibly have, especially if you live in an area where the water is particularly hard, because brass will not corrode. Also, brass is nearly as pressure-resistant as galvanized steel. Efficient as it is in plumbing systems, though, brass has become so expensive that practically no one can afford to install it anymore. Another liability is that brass piping requires a dielectric coupling or union when it is being connected to galvanized steel; the coupling is needed to prevent corrosion from setting in around the joint. (Neither copper nor plastic connections to a brass system require any special couplings.)

The biggest drawback to both brass and galvanized steel piping concerns their fittings: both metals must be connected with threaded fittings, and in order to cut the threads you must have a stock-and-die set. A stock and die can be purchased anywhere that plumbing hardware is sold, but a set can cost as much as $300. Of course, you can rent a stock and die from many plumbing suppliers and rental agencies, but using it is not easy—especially if you're inexperienced. For plumbing work that requires a stock and die, you might best hire a professional. If you decide to cut the threads yourself, use the stock and die carefully as follows:

Using a Stock and Die

1. Cut the pipe to its proper length, using a hacksaw and miter box or a pipe cutter.
2. Insert a pipe reamer in the cut end of the pipe and rotate it clockwise to scrape off all burrs. The outside of the pipe can be smoothed with a file.
3. The end of the pipe should be approximately 12″ beyond the jaws of your vise. Lock the proper-sized die for the pipe you are threading into the stock.

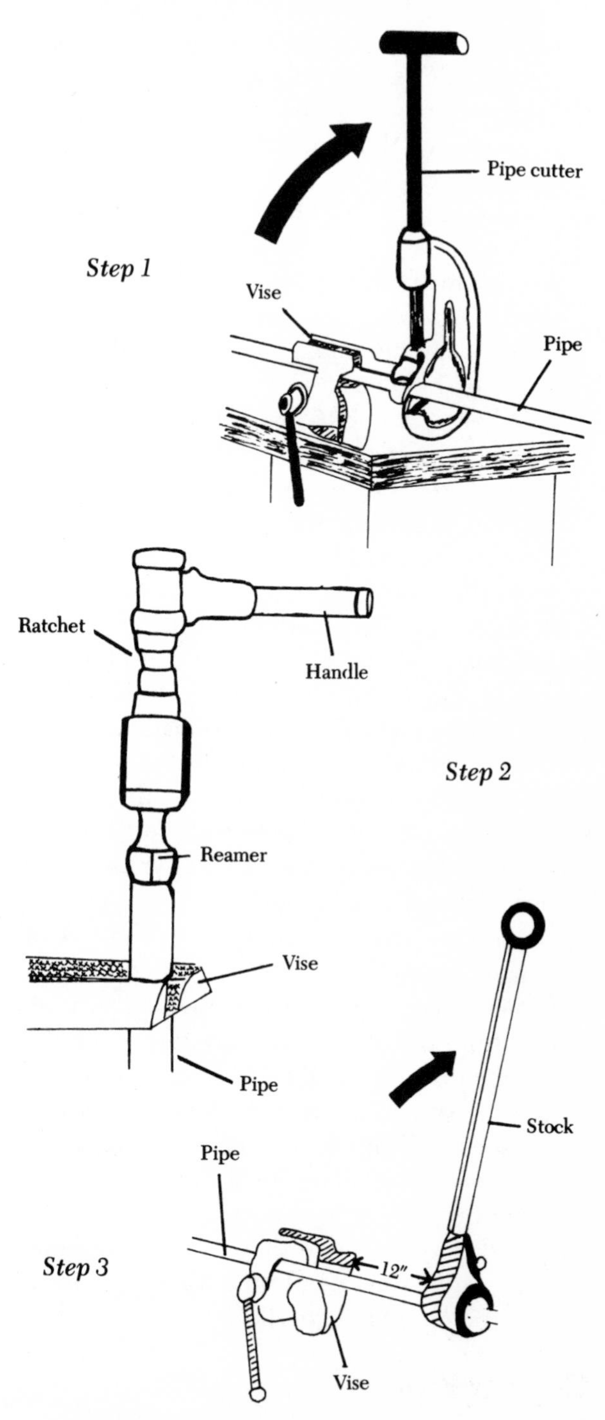

4. Slide the die over the end of the pipe. You will have to rotate it slightly until it grips the metal.

5. Turn the stock clockwise one complete turn, then stop and squirt cutting oil on the pipe through the oil ports in the side of the die. If the die binds at any time, rotate it counterclockwise about a quarter turn to allow any metal filings to fall out of the threads.
6. Continue turning the stock, pausing after each complete rotation to apply more cutting oil. Without oil, you risk blunting the die or breaking the threads on the pipe. When about 1/16″ of the pipe end protrudes past the face of the die, you are finished. Rotate the stock counterclockwise until it comes free of the pipe.
7. Wipe the threads with a soft cloth to clean it of any metal particles.

You can often purchase precut, threaded lengths of brass or galvanized steel pipe at plumbing supply outlets in lengths of 1″ to 12″, so you may be able to complete a small plumbing repair or add to your existing system without doing work that requires a stock and die. When buying precut lengths of pipe and their fittings, remember that measurements are critical and demand several precautions:

1. Measure the distance between the face of each fitting.
2. Insert your ruler in the opening of each fitting and determine how much of the pipe threading will nest in them.
3. Add both fitting socket measurements to the overall pipe length. Be sure to measure every socket in every fitting. There are considerable differences among sockets, and if you assume they are all the same depth you may end up with a pipe measurement that is short.

Copper Pipe

Until the advent of plastic, copper was considered the most versatile of the metal pipes and the easiest to work with. Consequently, many (if not most) homes in America have water supply systems constructed of copper pipe and tubing. Copper is light and durable, it resists the mineral deposits found in most of the water throughout the United States, and it is manufactured in both rigid and flexible form in diameters that permit its use in both DWV and water supply systems. It can be joined in three ways: by use of flare fittings, which are relatively expensive; with a compression fitting and ferrule; or by sweat-soldering, which requires the use of a $15 propane torch kit—and a few practice sessions with it.

Copper pipe is sold as tubing (soft tempered) and rigid pipe (hard tempered). The tubing is as good at bending around corners as plastic is, while the rigid pipe is excellent for long runs. The outside diameter of copper is generally ⅛″ larger than its nominal diameter, although the thickness of the copper can vary. The most common diameters found in residences are ¾″, ½″, and ⅜″ in the water supply lines, and 1½″, 2″, and 3″ in the DWV.

SOLDERING COPPER

The only way you can join the larger-diameter rigid copper pipe used in DWV systems is by sweat-soldering. With copper tubing or small-diameter rigid pipe you have the alternative of flare joints.

In order to sweat-solder, the metal must be heated enough so that it can melt the solder, and to do this you need a propane torch. The propane comes in a small metal canister which has a threaded nozzle to which you screw on a flamehead. If you are soldering large-diameter pipe (1½″-3″), one propane torch will not deliver enough heat to properly warm the metal; you must use two propane torches simultaneously, or a blow torch. No matter how you are heating the copper, the procedure for sweat-soldering is always the same:

1. Shine the end of the pipe and the inside of the fitting socket with steel wool or sandpaper. The pipe end should be cleaned about a half inch beyond the point that will enter the fitting so that the solder can flow easily around the joint.
2. Liberally coat all of the shined metal around the pipe and inside the fitting with a non-acidic paste soldering flux. The flux will prevent the metal from oxidizing when you heat it and allow the solder to adhere to the copper properly.
3. Assemble the fitting on the pipe and twist it until it is facing in the proper direction.
4. Open the control knob on the propane torch and light the torch with a match. Be careful where you aim the torch: you cannot afford to set fire to your house while all the water is turned off.
5. Heat the pipe and fitting by moving your torch flame over the metal. Keep the flame at right angles to the angles to the joint you want to solder, and constantly rotate around the pipe.

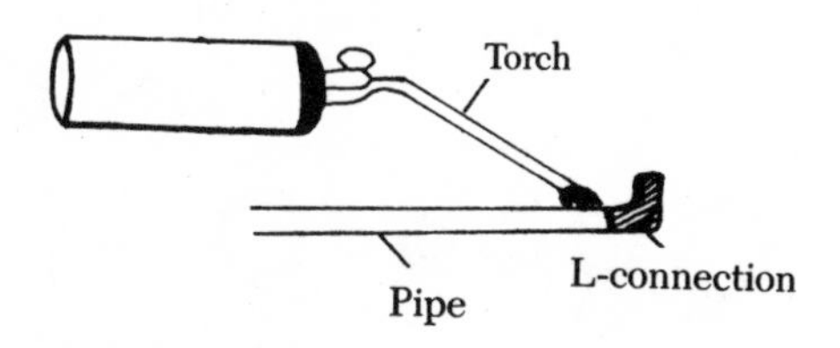

Keep the flame moving around the pipe.

6. Solder comes rolled around a spool and looks like a piece of aluminum wire. Unroll part of it and hold it against the copper. When the metal is hot enough to melt it, press the end of the solder against the edge of the fitting. The solder will slide around the pipe or actually be sucked around it, even if you are working vertically, by capillary action.

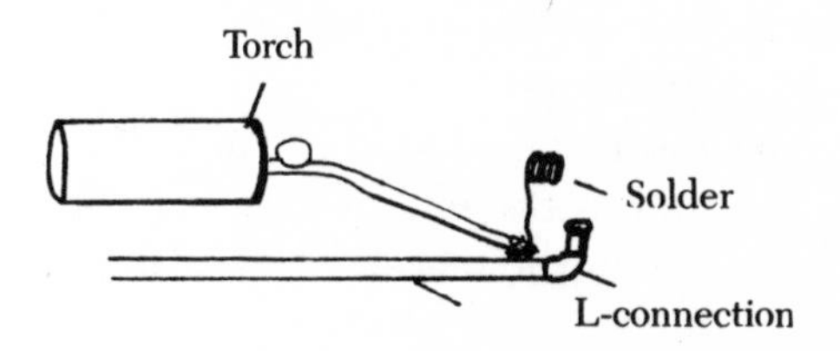

When the metal is hot enough to melt the solder, hold the solder against the pipe.

7. Inspect the joint. Any areas that do not seem to be filled with solder can be given an extra daub.
8. Remove the flame of your torch. Not immediately, but while the solder is still warm and fluid, wipe it with a soft cloth to get rid of any excess flux and solder. By the time the pipe has cooled enough to be handled, the joint will be watertight.

MAKING FLARE JOINTS

Flare joints require the use of flare fittings, which are considerably more expensive than a standard soldered fitting. Consequently, if there are a great many connections to be made, soldering is usually preferably over the flare assembly.

There are two kinds of flaring tools, the impact type and the screw type. Neither is very expensive or difficult to use.

Using an Impact-Type Flaring Tool

Flare fittings have a conical end that is threaded. The cone nestles into the flared end of the pipe, and the two pieces are held together in a watertight seal by a nut that fits over the connection and is tightened to the threaded portion of the fitting. Each fitting comes with a flaring nut attached to each of its ports, but you must flare the end of the pipe.

1. Remove the flaring nut from the fitting and slide it over the end of the tube so that its threads face the end of the pipe.
2. Secure the end of the pipe in a vise or similar grip.
3. Insert the pointed end of the impact flaring tool in the end of the pipe. Be careful to center the tool and keep it aligned with the pipe.
4. Tap the end of the tool with a hammer. You will only need two or three light taps to drive the tool deeper into the pipe and flare its rim.

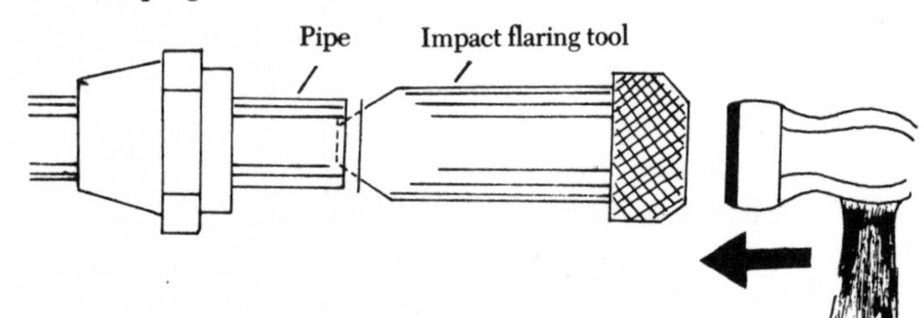

Lightly tap the impact flaring tool with a hammer.

5. Remove the flaring tool and insert the conical end of the fitting against the flare. Then tighten the flaring nut on the fitting. You do not need to wrap the fitting threads with plastic tape or plumber's putty.

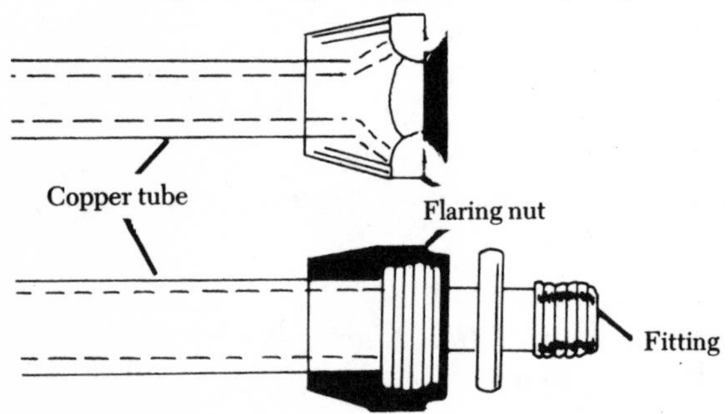

Hold the fitting against the flared end of the pipe and tighten the flaring nut on the threads of the fitting.

Using a Screw-Type Flaring Tool

A screw-type flaring tool is somewhat more reliable than the impact tools because its vise locks the end of the pipe at a precise right angle with the screw.

1. Place the flaring nut on the end of the pipe.
2. Insert the pipe in the proper diameter hole in the vise block and lock the jaws around it. The pipe should extend about ⅜″ beyond the block.
3. Center the yoke on the block and insert the cone on the end of the screw handle in the pipe.
4. Tighten the screw handle until the end of the pipe has completely flattened against the block.
5. Remove the pipe from the yoke and assemble the fitting to the flaring nut. When tightening the nut, place one pipe wrench on the nut and another one on the fitting.

Using Compression Fittings

Compression fittings also involve a nut which screws to the fitting, plus a brass ferrule. Most often, you will encounter compression fittings on the ⅜″ tubing used to connect the shutoff valves in your water supply lines to faucets and toilet flush valves. Once the brass ferrule has

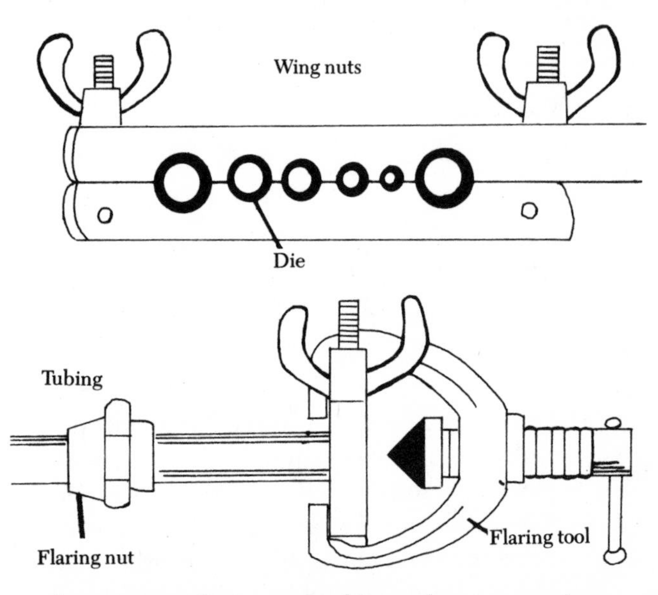

A screw-type flaring tool achieves the same result as an impact-type flaring tool.

been squeezed around the tubing, it is extremely difficult to remove; even if you do get it off the pipe, the ferrule will probably be bent out of shape and have to be discarded. Replacement ferrules can be purchased singly or in packages of assorted sizes.

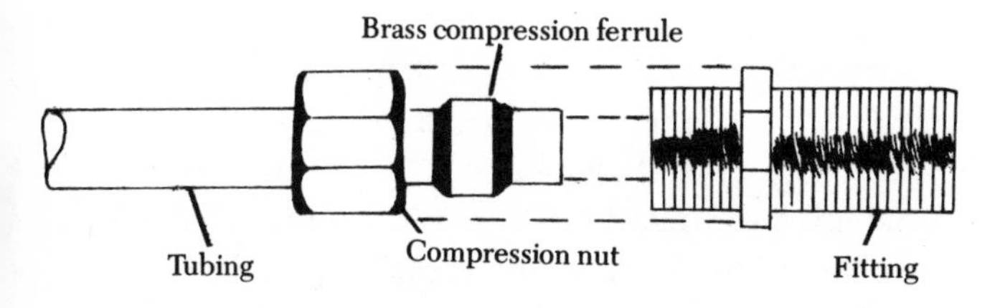

The compression ferrule is usually brass or plastic.

1. Slide the compression nut over the pipe, with its threads facing the direction of the fitting.
2. Slide the brass ferrule onto the end of the tubing.
3. Insert the tubing in the fitting and tighten the nut to the fitting. By tightening the nut you will also squash the ferrule around the tube, creating a watertight seal.

CUTTING COPPER

Copper pipe and tubing can be cut with a hacksaw, if necessary. It is important to have a very straight cut in copper pipe, so if you are using a hacksaw, place the pipe in a miter box when you saw it. A much neater and easier way of cutting is to use a pipe (or tubing) cutter. This is a C-shaped tool with a cutting wheel that is tightened against the surface of the metal via a knob.

1. Measure and mark the pipe for cutting.
2. Place the cutter around the pipe, with the cutting wheel on your mark.
3. Tighten the cutter knob until the wheel is snugly against the metal.
4. Rotate the cutter around the pipe until it moves freely.
5. Tighten the cutter knob until the cutter cannot move easily.
6. Rotate the cutter around the pipe.
7. Continue rotating and tightening the cutter until the pipe is severed.
8. Remove all burrs from the inside and outside of the cut edge with a fine file or sandpaper.

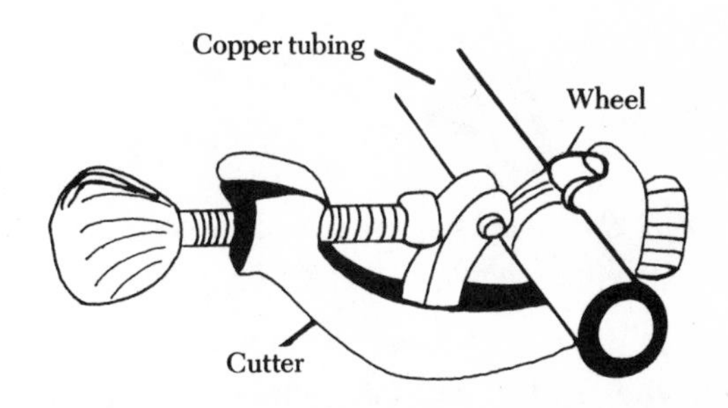

Tubing and pipe cutters are tightened against the metal and then rotated in either direction to cut the pipe.

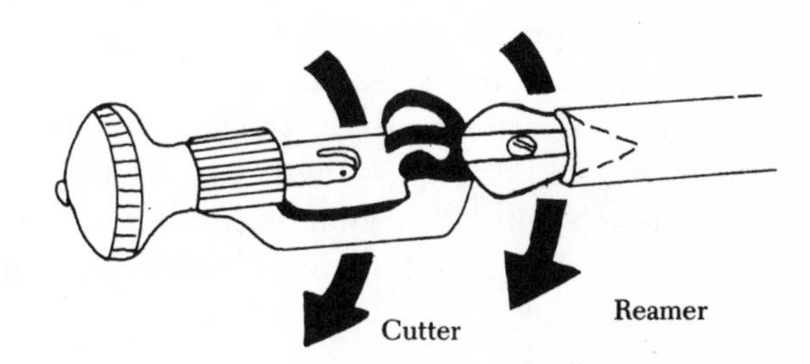

The reamer is a pointed metal blade attached to the side of the cutter.

REMOVING COPPER PIPE AND FITTINGS

If you are working with copper piping that is already installed, your first task will be to remove part of the pipe or some of its fittings. Compression or flare fittings are loosened by applying one pipe wrench to the fitting to steady the assembly, and then undoing the nut with a second wrench. If you are removing a section of pipe, cut it with either a hacksaw or a pipe cutter.

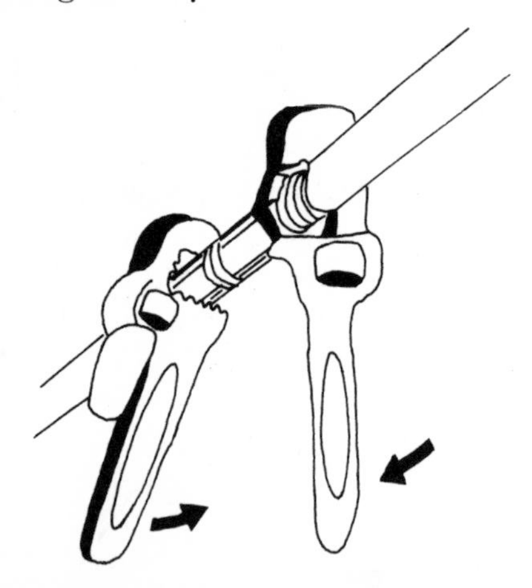

Be sure the mouths of the wrenches are facing the direction in which you will move the wrench.

To remove a soldered fitting, you must first melt the solder with the flame of a propane torch.

1. If you are loosening only one of the joints in the fitting, wrap the other joint with wet rags to keep its solder from melting.
2. Apply the flame from a propane torch to the joint to be loosened.
3. When the solder around the joint begins to melt, tap the pipe with a hammer or wrench until the joint pops free of the fitting. Be careful not to touch the pipe; it is very hot.

REPAIRS TO COPPER PIPE AND TUBING

The surest method of repairing a broken solder joint is to melt the solder and remove the fitting, clean the pipe ends and the fitting (or replace the fitting), and install the union again.

Epoxy cement can be used to repair minor leaks in a joint. Be warned, though, that a joint which has been repaired with epoxy will be extremely difficult to dismantle in the future. If further repairs need to be done at a later date, you may have to cut the fitting off the pipe.

When leakage appears at a flare or compression fitting, tighten the nut on the fitting until the leak stops. If tightening fails, the fitting may have cracked or its threads may have been stripped. It should be replaced.

Cast Iron Pipe

Cast iron is most frequently used in the construction of soil stacks and house drains, although galvanized steel, brass, and now plastic are also used. You can buy cast iron pipe in 2″, 3″, and 4″ diameters, and in 5- or 10-foot lengths. While cast iron will provide a good 40 or more years of trouble-free service, it is awkward to work with and, of course, expensive. Most local plumbing codes now permit brass or galvanized steel pipe that is either 3″ or 4″ in diameter instead of cast iron, as well as plastic pipe. It is also possible to adapt a plastic drain line to cast iron by using specially threaded fittings that will accept a coupling.

There are two forms of cast iron pipe. The oldest version is known as **bell-and-spigot** or

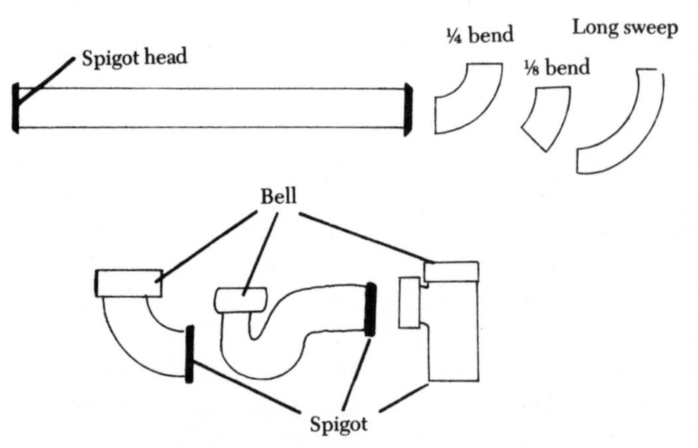

Cast iron hub pipe has a bell at one end and a spigot at the other; hubless pipe is nothing more than a straight piece of pipe, although some versions have positioning lugs around their ends.

hub pipe, which has a bell-shaped end designed to accept the ridge around the spigot end of each length of pipe. The newer version of cast iron is known as **hubless pipe** and is nothing more than a straight length of pipe that is connected with a neoprene sleeve.

MEASURING AND CUTTING CAST IRON PIPE

When measuring the length of cast iron hub pipe, always remember to allow for the amount of spigot end that will be in the hub:

2″ pipe—allow 2½″
3″ pipe—allow 2¾″
4″ pipe—allow 3″

Hubless pipe needs no allowances made for its fittings.

Both hub and hubless cast iron pipe are cut in the same way:

1. Measure and mark the pipe. Draw your cutting mark all the way around the pipe.
2. Cut a 1/16″-deep score all the way around the pipe with a hacksaw.
3. Insert the blade of a cold chisel in the groove and tap it lightly. Rotate the pipe and tap completely around the score in the pipe.

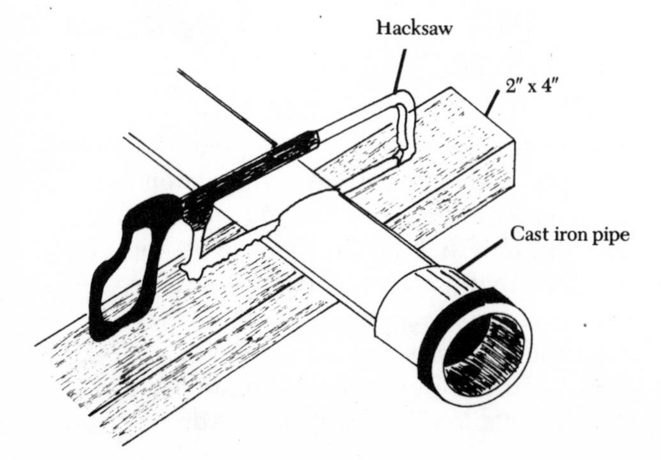

4. Continue tapping the groove until the pipe produces a dead, hollow sound.
5. Strike the pipe with a hammer, hitting it on the waste side of the groove. The waste portion will break off.

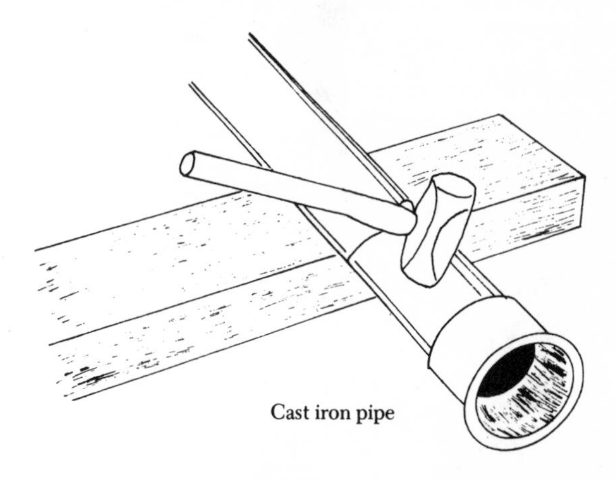

JOINING HUB PIPE (BELL AND SPIGOT)

Bell-and-spigot cast iron pipe must be joined with molten lead and oakum, which requires a burner to heat the lead, and a caulking tool, oakum, a ladle, and a joint runner. You can rent most of this equipment or, if you have a considerable amount of cast iron work to do, buy it at plumbing supply outlets.

When joining hub pipe, the hub must always point upward toward the direction of the water flow, so that waste cannot get caught in the joint.

1. Insert the spigot end of the pipe in the hub so that it stands snugly and evenly.
2. Use a caulking tool to cram lengths of oakum into the hub, around the spigot. Fill the hub to within one inch of its rim.
3. Melt enough lead to fill the hub with one ladleful.
4. Pour the lead over the oakum, filling the bell. Allow the lead to cool.
5. When the lead has cooled, pack down any excess with a cold chisel or the caulking tool. *(see illus. p. 92)*

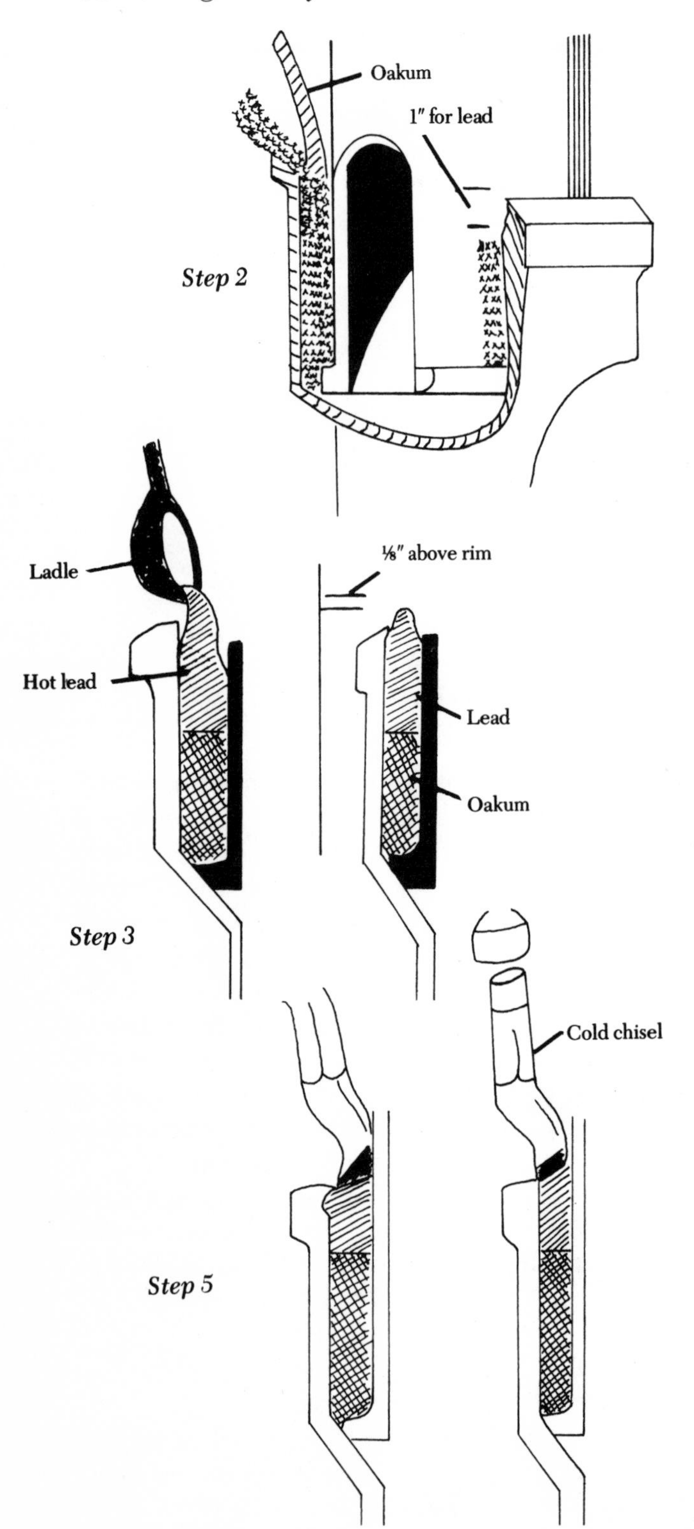

If the assembly is parallel with the floor, you must clamp a joint runner around the top of the bell to hold the lead in place while it is hardening.

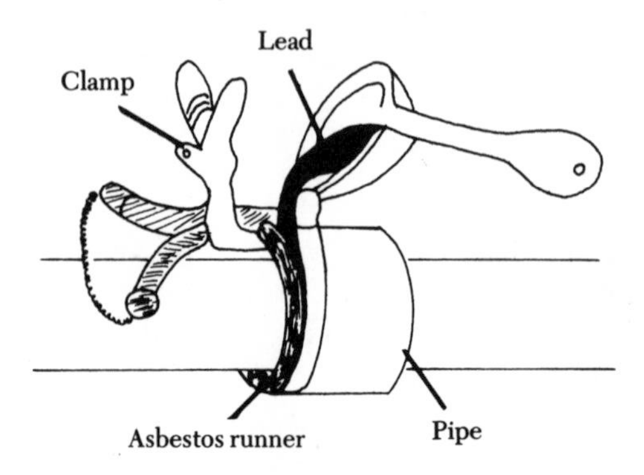

A joint runner is clamped around horizontal pipes to hold the lead in place until it hardens.

JOINING HUBLESS CAST IRON PIPE

Hubless cast iron pipe is joined with a neoprene sleeve and gasket, which requires the use of a screwdriver.

1. Place the gasket on one of the pipes to be joined.

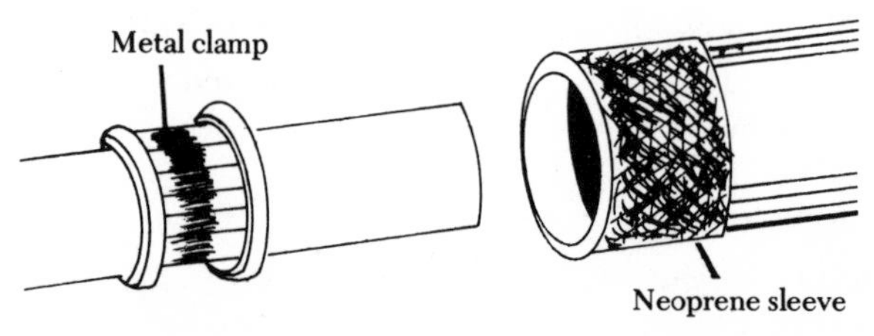

2. Slide the neoprene sleeve onto the other pipe. The sleeve has a ridge around the center of its inside, so you will be able to get only half of the sleeve onto the pipe before it bottoms against the ridge.
3. Insert the second pipe into the sleeve.
4. Slide the gasket over the sleeve.
5. Wrap two pipe clamps around the gasket and tighten them with a screwdriver.

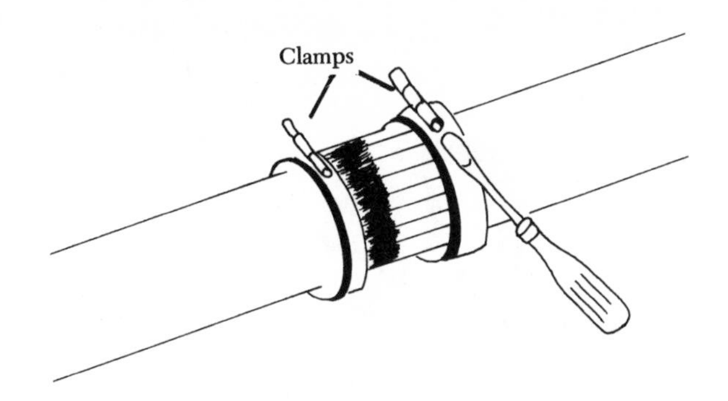

SUPPORTING CAST IRON PIPE

Cast iron pipe is extremely heavy and must be fully supported whenever it makes a horizontal run. It can stand vertically from your basement to the roof without any great problems, and the various drain lines that connect to it will serve to steady the entire stack. But with a horizontal run, cast iron pipe should not only be suspended from the joists by metal pipe hangers at intervals of 4 feet, but each joint should be resting on a length of 2″ × 4″ board set vertically underneath.

Plastic Pipe

Check your local and regional plumbing codes before going to work with plastic pipe and tubing.

There are five different types of plastic pipe. Rather than trying to remember the name and purpose of each version, explain the job you are doing to the salesman at the plumbing supply outlet and ask him for the proper type of plastic pipe. The different types are:

Polyethylene (PE)—PE is sold as flexible coils in ½″, ¾″, and 1″ diameters, and is black. It can only be used for transporting cold water, such as to a toilet or from a well. It can withstand pressure up to 160 lbs. p.s.i.

Polybutylene (PB)—This is beige or black in color and is manufactured in flexible 25-and 100-foot coils with ½″, ⅜″, and ¾″ diameters. It can withstand 100 lbs. p.s.i. and 180°F temperatures, and is therefore suitable for both hot and cold water supply lines.

Chlorinated polyvinyl chloride (CPVC)—White in color and rigid, CPVC is also able to withstand 100 lbs. p.s.i. pressure and 180°F of temperature, and can be used as an alternative to Polybutylene. It is sold in ¾″ and ½″ diameters in 10-foot lengths.

Polyvinyl chloride (PVC)—PVC is rigid and usually used for drain lines, although it can also carry cold water. It is white, gray, or beige, and particularly resistant to chemicals.

Acrylonitrile-butadiene-styrene (ABS)—ABS is black in color and used strictly for DWV lines. Like PVC, it comes in 10-foot lengths and a variety of diameters including 1¼″, 1½″, and 2″, as well as 3″ and 4″ for use in soil stacks and house drains.

Plastic Pipe Fittings

Manufacturers of plastic pipe have been hard at work developing all conceivable types and sizes of fittings, which allow you to do virtually all plumbing work with plastic pipes—including linking them to an existing system of metal pipes. While most plastic fittings are solvent-welded to the pipe they unite, there are also threaded couplings that can be connected to brass or galvanized steel, as well as fittings with serrated ends that fit tightly inside the pipe and are held in place with a stainless steel, non-corrosive pipe clamp.

HOW TO MEASURE PLASTIC PIPE

1. Any length of plastic pipe incorporated in a plumbing system will begin and end with a fitting of some sort. Measure not only the exact length of the pipe, but add to that length the inch or two needed to enter each fitting.

2. Plastic fittings all have ridges inside their sockets to stop the pipe. Place the end of a ruler in the fitting until it touches the ridge, and then read the depth of the fitting socket. Add the socket measurement to the projected length of the pipe.
3. If there are to be any fittings in the middle of the pipe run—a branch-T, for instance—be sure to allow for the length of that fitting in your measurements.

HOW TO CUT PLASTIC PIPE

1. Measure and mark the pipe.
2. Place the pipe in a miter box.
3. Cut through the pipe with a hacksaw.
4. Using a fine-grain file, remove any burrs or rough edges around the inside of the cut end of the pipe.
5. Using fine- or medium-grit sandpaper, sand off any burrs or rough edges on the outside of the cut.

Actually, you don't need to have a hacksaw to cut through plastic pipe. You could use any fine-toothed carpenter's saw, or even a kitchen knife. If you have a considerable amount of cutting to do, invest in a plastic-pipe cutter, which looks and functions like the standard pipe or tubing cutter (see page 121). The standard metal cutter, however, will tend to collapse the plastic, so be careful not to put too much pressure on the pipe.

The important point to remember when cutting plastic is to make the cut as straight as you possibly can so that the pipe will be completely seated against the ridge inside its fitting and form a watertight seal.

HOW TO SOLVENT-WELD PLASTIC PIPE

You can now buy all-purpose cements which will join any of the plastic pipes. But don't be stingy about it: buy the highest grade you can find, preferably one that is made by the manufacturer of the pipe you are using.

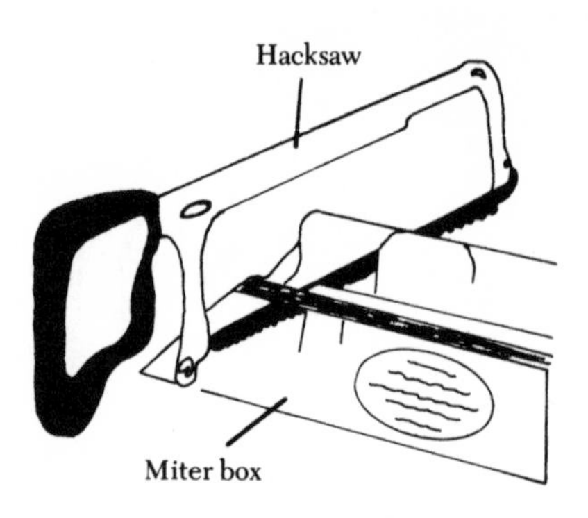

Use a miter box when cutting plastic pipe to guarantee a right-angle cut.

Make absolutely certain that the pipes and fittings you are about to assemble are the correct length, and that you know the proper positioning of the fittings. You will have about 60 seconds after applying the cement to make your connections. After that, there is no way you can correct a mistake other than to cut off the fitting and start all over again. This is the one drawback to plastic pipe: mistakes cannot be rectified. Your work must be perfect the first time.

If you do err and have to cut off a fitting, you will end up with a length of pipe that is too short. You can either replace the entire pipe with new fittings, or you can cut off a little more of it and insert a coupling with a small length of pipe in its free end to accept a new fitting. Either way, the fitting cannot be reused.

1. Check the cut end of the pipe to be sure there are no burrs or rough areas.
2. Wipe both the end of the pipe and the socket of the fitting with a plastic-pipe cleaner to remove all dirt and grease. You may think the pipe is clean when you touch it, but the cleaner will take off a surprising amount of grime (and oils from your fingers) that would otherwise inhibit the action of the cement.
3. Allow the cleaner to evaporate on the sur-

faces to be joined; the pieces must be absolutely dry.

4. You can apply the cement with the applicator that is attached to the cap of the can, or you can use a narrow paint brush. Liberally coat the inside of the socket in the fitting. Apply a heavy coat of cement around the outside of the pipe end. Be very careful to cover all of the surfaces; it is better to apply too much cement than not enough.
5. You don't have much time. Push the pipe into the fitting at about 5° off its final position, then twist the fitting until it is seated properly.
6. Hold the joint together for 10 seconds. The solvent will set in 30 seconds, and will be able to withstand water pressure within an hour. But allow it a minimum of 16 hours before you pressure-test it.

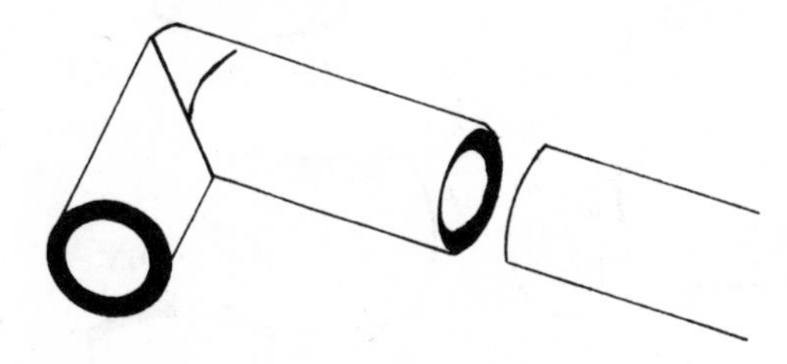

Hold the fitting and pipe together for 10 seconds to allow the solvent-weld to set.

Joining Plastic Pipe to Metal

When you are joining plastic pipe to any of the metals, you can always find an adapter or transition fitting. Some are threaded at one end, others are made of metal and have a plastic insert that can be cemented to the plastic pipe. PE pipe is joined to any threaded pipe with a threaded or serrated adapter that screws to the metal pipe at one end and pushed into the plastic pipe at the other. The assembly is then secured with a worm-drive pipe clamp.

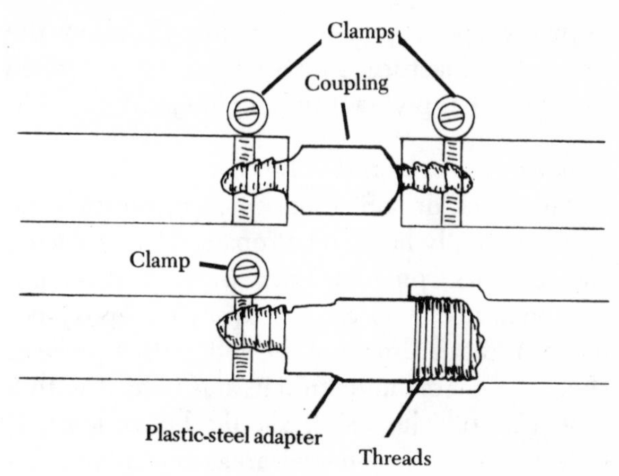

Plastic pipe is connected to metal with plastic-to-metal adapters.

REPAIRS TO PLASTIC PIPE

If you make a mistake assembling plastic pipe and find that you must cut the fitting out of the pipe and replace it, you can retrieve the inches you have lost in the pipe run by inserting a sleeve or coupling. The procedure is this:

1. Remove the damaged section by cutting it off the pipe with a hacksaw.
2. Measure a new section of pipe and cut it. Be sure to make allowances in your measurements for joining a new fitting at one end and a sleeve at the other.
3. Cement the fitting and sleeve to the new piece of pipe.
4. Install the new section to the damaged pipe.

Repairing Leaks at a Joint

A properly assembled fitting will have an even bead of cement around its base. A joint that does not have enough cement may not be watertight, which means it will leak. If you assemble a fitting and do not have an unbroken bead of cement, use more cement in your next connections. Simply smearing more cement around the joint will not solve the problem; the cement cannot be used as caulking. But you can block the leak by applying epoxy cement. Be sure to follow the epoxy manufacturer's instructions exactly when mixing and applying it to the pipe joint.

Repairing Cracks and Splits

Any crack or split that occurs in plastic pipe will eventually have to be repaired by replacing the section of pipe with new pipe. As a temporary measure, you can try applying epoxy cement. You can also wrap the split with a piece of sheet plastic or inner tubing and hold it with a pipe clamp. The patch should be at least 3″ longer than the damaged area, but should be slightly narrower than the circumference of the pipe. Alternatively, pipe clamps can also be used around the damaged area.

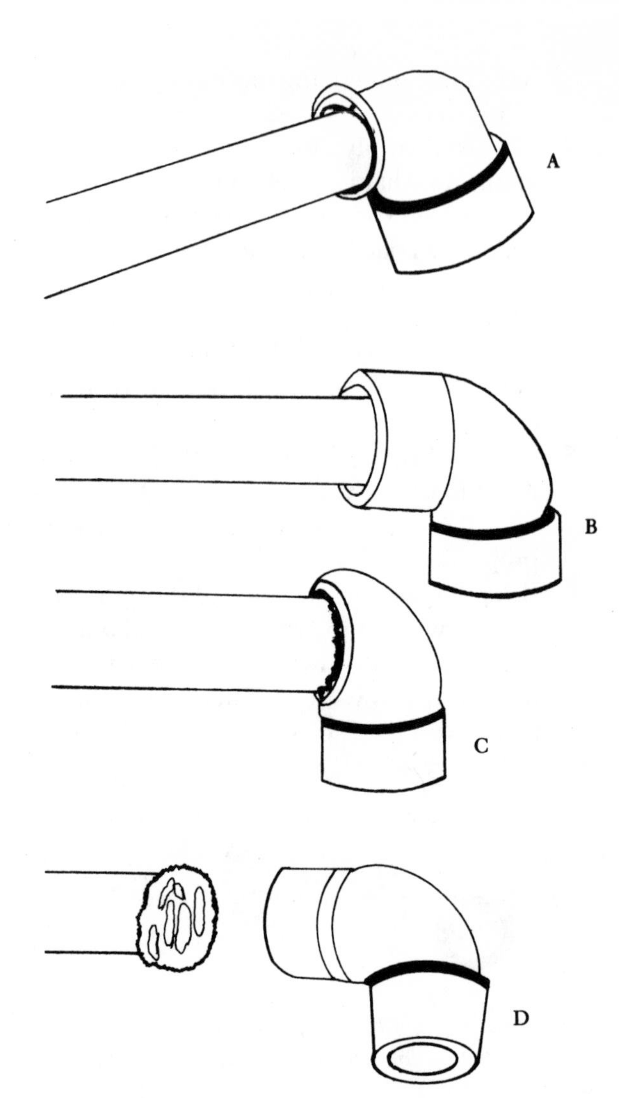

A correct joint has an even bead of cement around its edge (A). If there is no bead, the joint could leak (B). When the bead is uneven, the joint could leak; use more cement next time (C). Too much cement can clog the fitting socket and pipe (D).

CHAPTER 6:

Plumbing a New Bathroom

THERE IS SOMETHING infectious about doing your own plumbing. For one thing, every time you stop the drip in a faucet you can count about $20 saved that might have been spent hiring a plumber. By the time you have installed a new kitchen sink or replaced a hissing flush valve in a toilet, you may feel more confident about taking on a whole plumbing job, such as renovating an old bathroom or building a new one. And if you elect to use plastic pipe, the job can even be fun.

Installing a new bathroom involves practically every problem and consideration that any professional plumber ever faces. So in a sense, adding a second bathroom in your home becomes a test of all your plumbing skills. It entails not just all of the plumbing techniques, but some basic carpentry. And if you happen to surface the floor and walls with ceramic tile buried in cement, you will be doing masonry work as well.

Planning a Plumbing System

The cost of pipe and fittings is high, even if you are using plastic. So when you are planning any major plumbing alterations or additions, always think in terms of every pipe going as directly as possible to its destination with the fewest joints possible. Although you could use galvanized steel, copper, brass or cast iron pipe to construct a three-fixture bathroom (tub-shower, sink, toilet), it is less expensive, easier, and more lasting to use PVC plastic pipe for your entire drain-waste-vent system and CPVC and PB for the water supply lines.

You should have a plan before you begin work.

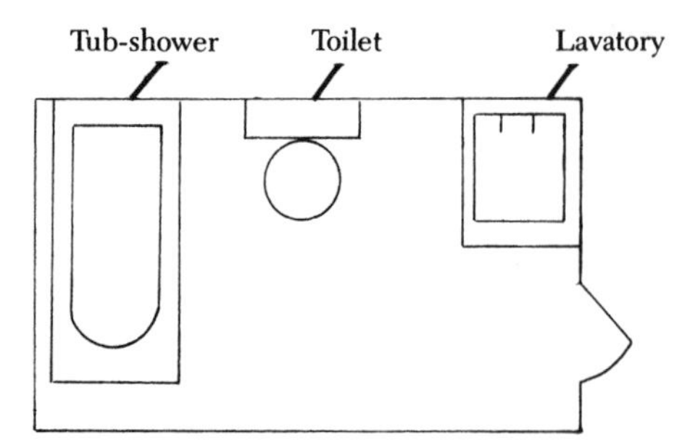

Typical three-fixture bathroom layout

The plan can be a simple, rough sketch of the bathroom and the DWV- and water-supply-line connections. Or it can be an elaborate architectural blueprint that includes every fitting and the length of every pipe run. Besides being a roadmap for plumbing a new bathroom, your plan can save you a number of trips to the nearest plumbing supply store for unanticipated fittings. You may still need to make an occasional shopping run, even with very careful and complete planning, but the trips will be fewer.

PLUMBING CODES

The United States of America abounds in plumbing codes. There are local codes in every municipality: state codes, city codes, county codes, and national codes. The codes are not there to harrass you, although sometimes they make some very expensive or time-consuming demands. The purpose of all the codes is to establish and maintain sanitary plumbing systems that present no health hazard whatsoever. Primarily, the codes exist to ensure that there are no cross-connections between the DWV and water supply lines that might allow waste to back up into the potable water in your home.

The codes are excellent guides to efficient, sanitary plumbing systems, and can be very useful. If you are only doing occasional minor repair work, you will not be required by your local building or health department to have a permit. But if you plan to add a bathroom in your house, you are expected by most municipalities to register your plans and obtain a building permit from the buildings department. The permit will be granted to you for a small fee, but with it comes the requirement that your work be officially inspected at various stages. Bothersome as this may be, passing those inspections will assure you that your new plumbing system is both safe and durable.

Bill of Materials

After you have worked out your plans, make up a bill of materials, or shopping list. If you are using plastic pipe, be aware that some manufacturers have assembled whole kits of all the fittings needed to plumb a basic bathroom. This kind of kit just might be ideal for your plumbing needs, although you might get into special situations beyond the bathroom that require extra fittings or pipe.

The essential Bill of Materials for a standard three-fixture bathroom will include approximately these plumbing materials (presuming you are using PVC in the drain-waste-vent system, and CPVC and PB for your water supply system):

DWV MATERIALS

Quantity	*Size*	*Description*
1	3″ × 1½″	Special Waste-and-Vent Fitting*
1	3″ × 4″	Toilet flange
1	3″	Coupling
1	1½″	90° Elbow
1	3″	90° Elbow
1	3″	Y-fitting
1	3″ × 3″ × 1½″	Reducing-T
1	3″	Cleanout and plug
1	1½″	45° Street elbow
1	3″	45° Street elbow
1	1½″ × 1¼″	Male trap adapter
1	1½″	Trap adapter
1	1½″	45° Street elbow

1	1½″	P-trap with cleanout plug
1	3″ × 3″	2-piece roof flashing
1	10′	1½″ PVC pipe
2	10′	3″ PVC pipe**
1	3″	Adapter for connecting the 3″ drain lines to whatever metal pipe is in your existing drain line
1	Pint	PVC solvent-weld cement

* The Special Waste-and-Vent Fitting has several variations, all of which allow you to connect the toilet and also tap in 1½″ or 2″ drain lines from the tub and lavatory. If the SW&V is not used, you will need an ordinary sanitary-T fitting plus several other T's and Y's to accomplish the same connections.

** If you need to go up two stories to vent through the roof, buy a third 10-foot length of pipe.

WATER SUPPLY MATERIALS

Quantity	*Size*	*Description*
11	½″	CPVC/PB Adapters
7	½″	T's
4	½″	Pipe caps
2	½″	Unions for the tub/shower valve
1	½″	90° Elbow
3		Angle-stop riser-tube adapters
2	⅜″ × 12″	Riser tubes (to connect the faucets)
1	⅜″ × 12″	Toilet riser tube
3		Escutcheons
1	½″	Wing elbow
2		Line stop valves
1	½″	Male threads adapter
3		Angle fixture shutoff valves
25′ coil	½″	PB tubing
10′	½″	CPVC pipe
	¼ pint	PVC solvent cement
2		Line adapters to connect with your existing hot and cold water lines

Roughing-in

There are some hard rules that must be followed when you are cutting a path through the frame of your house for a plumbing system. The rules are fully supported by the laws of nature, so you can't ignore them.

Roughing-in consists of drilling holes, cutting notches, and otherwise making space between

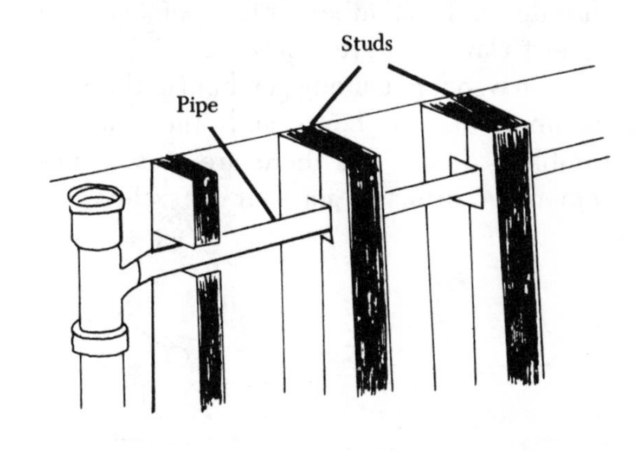

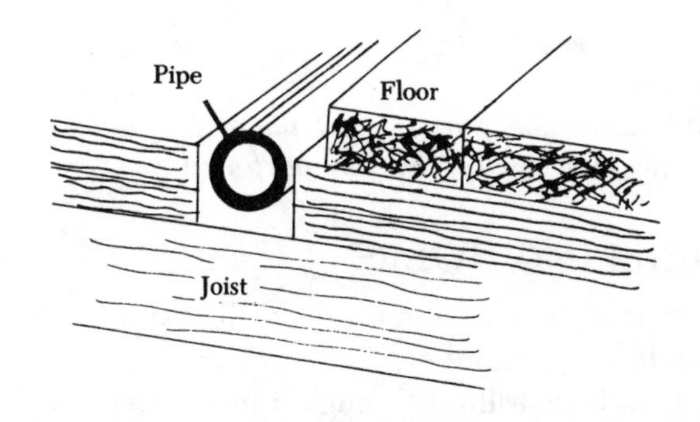

Roughing-in is making space for your pipes in the walls or floors.

and around the framing members inside the floors, walls, and ceilings of your house so that all of your plumbing lines can run directly and unobtrusively to and from the fixtures. To accomplish all this you will need an assortment of chisels, a utility handsaw, and a hammer. But probably the two most useful tools are a saber saw and an electric drill with a complement of spade bits.

Your pipes have to go over, under, or through the 2″ × 4″ studs and headers inside your walls, as well as the 2″ × 8″ or 2″ × 10″ joists that hold up the floors in your house. It is best to attach a pipe to the side of any framing member using pipe support brackets which are merely nailed or screwed into the wood. But there are times when you absolutely must go through a wooden member, which means you have to either drill a hole large enough for the pipe, or cut a notch in the edge of the member. At that point you are up against a law of nature: holes and notches tend to weaken wood; put enough of them in the framing members of your house and "the walls come tumbling down." So there are some rules to remember, primarily for your own safety.

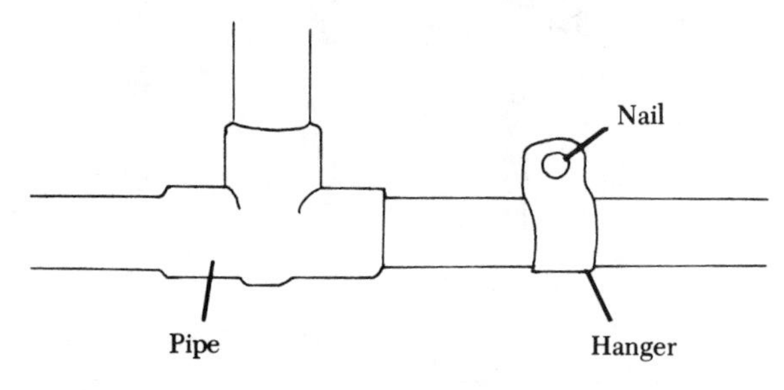

Whenever you can, support pipes with hangers, rather than notch or drill through framing members.

DRILLING HOLES

It is always preferable to drill, rather than notch, but remember:

1. When drilling through a joist (which will be 8″ or 10″ wide), the hole must be at least 2″ away from any edge of the wood.

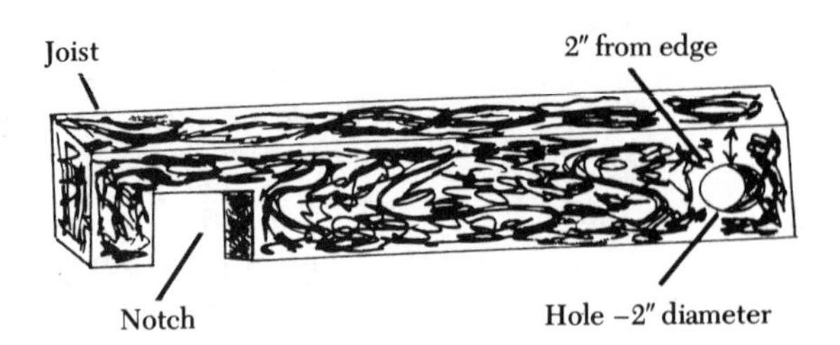

2. The diameter of the hole should not exceed one-quarter of the width of the wood, which will be 2″ or 2½″ for most joists.
3. Because studs are normally vertical members and the weight of the house pushes down on their length, they can withstand a drilled hole more readily than horizontal joists. Nevertheless, keep your holes as small and as far apart as possible.

NOTCHING STUDS

1. Studs should be notched no deeper than two thirds of the width of the wood. This means the notches in a 2″ × 4″ stud would be no deeper than 2½″.
2. Attach a strip of wood or steel bracket across the faces of each notch you make.
3. Never notch more than two adjacent studs. Try to leave at least two unnotched studs on either side of each notched pair.

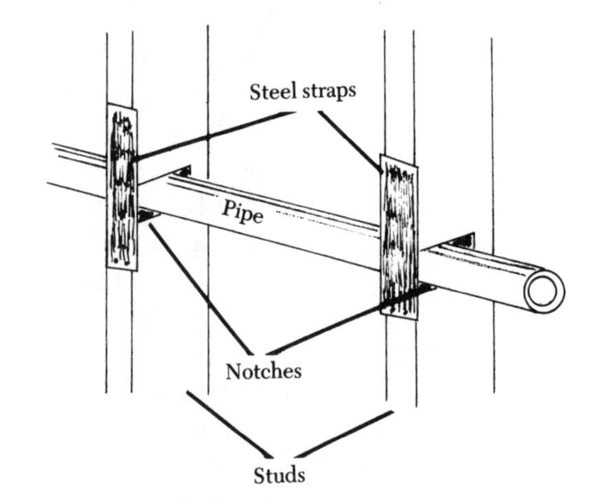

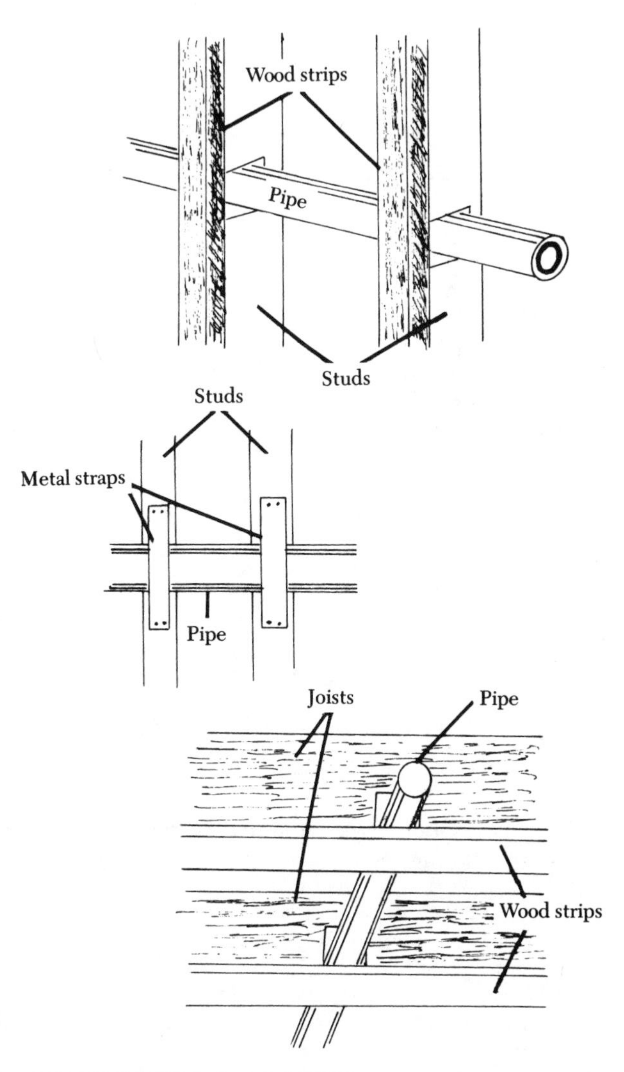

NOTCHING JOISTS

1. If you divide the length of a joist into four equal parts, all notches should be made only in the two end quarters.
2. Notches should be as shallow as possible and never deeper than one quarter of the width of the joists—that is, 2″ or 2½″ maximum.
3. Nail or screw a steel plate across the mouth of each notch after the pipe has been inserted in them. Or you can nail pieces of 2″ × 4″ or 2″ × 6″ lumber to either side of the notch for even greater strength.

How to Notch a Stud or Joist

1. Make two parallel saw cuts in the edge of the stud or joist to the depth of the notch and far enough apart to accept the pipe.
2. Place the blade of a chisel at right angles between the bottoms of the two cuts.
3. Hit the chisel with a hammer.

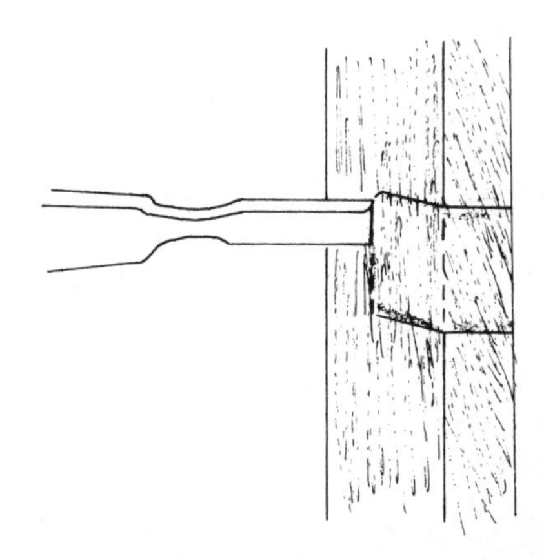

Plumbing the Drain-Waste-Vent System

When plumbing a new bathroom, the work always begins with the DWV system, which is organized around the toilet bend or Special Waste-and-Vent Fitting (SW&V). The toilet is held in place by the toilet flange, which should be installed in the floor 12″ from the finished wall. Your 12″ measurement can be taken from the face of the studs if the wall is not yet completed, but be sure to add the thickness of whatever wallboard and tiles will be attached to the studs.

The toilet can be installed anywhere, but it should be centered between floor joists. Bear in mind that the wall behind it must contain the soil stack, which will go down to the basement and connect to the building drain, as well as up to the roof.

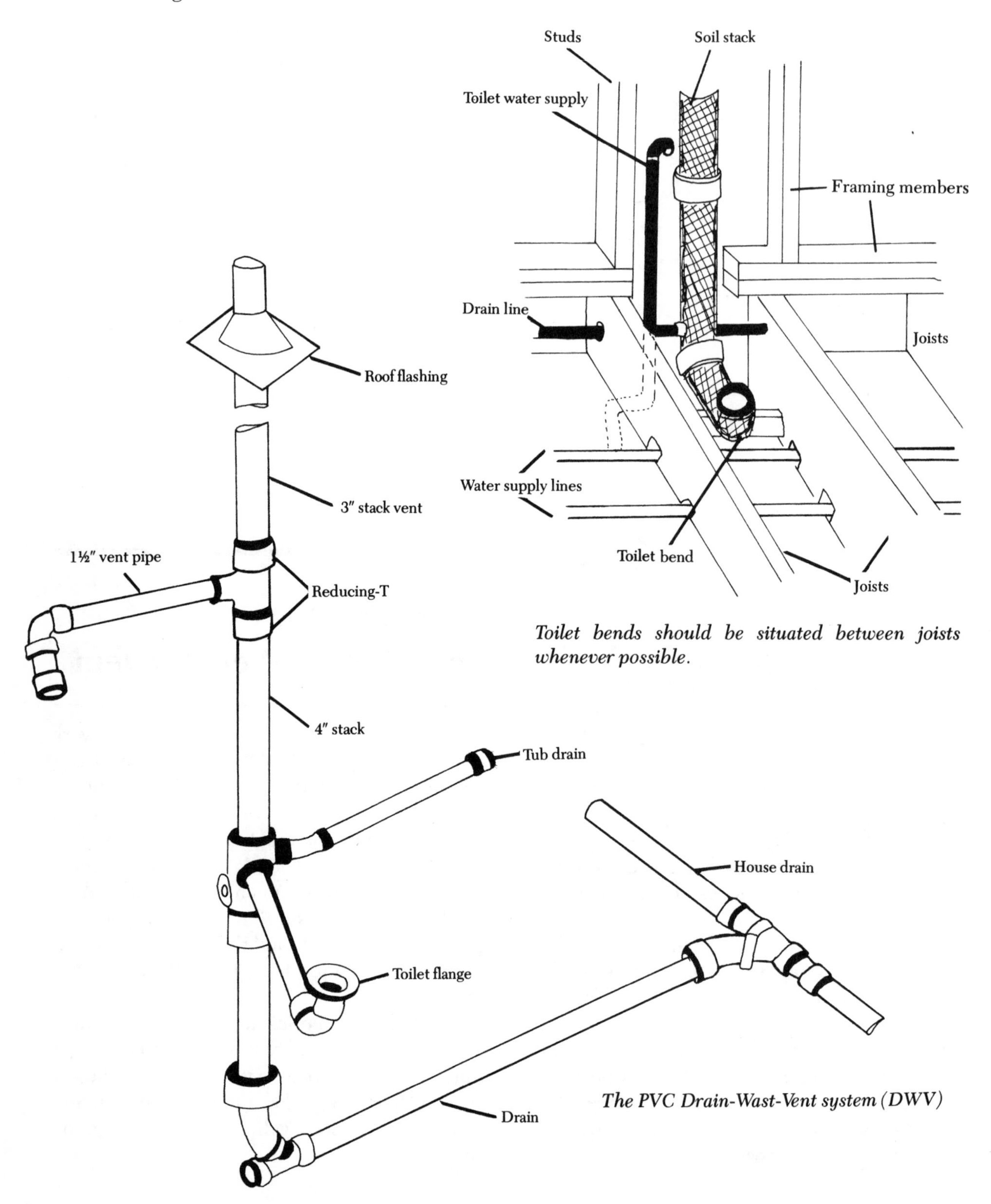

Toilet bends should be situated between joists whenever possible.

The PVC Drain-Wast-Vent system (DWV)

Installing the Special Waste-and-Vent Fitting (SW&V)

1. Measure 12″ out from the finished wall that will be behind the toilet and mark the center of the hole for the toilet flange. Draw a 12″-diameter (6″-radius) circle on the floor.
2. Drill a hole in the floor at the edge of the 12″ circle. Cut the circle with a saber saw.
3. Cement a 90° toilet elbow to the bottom of the flange and insert the flange in the hole. It should rest evenly and solidly against the floor, with the elbow facing the place where you intend to position the SW&V fitting.

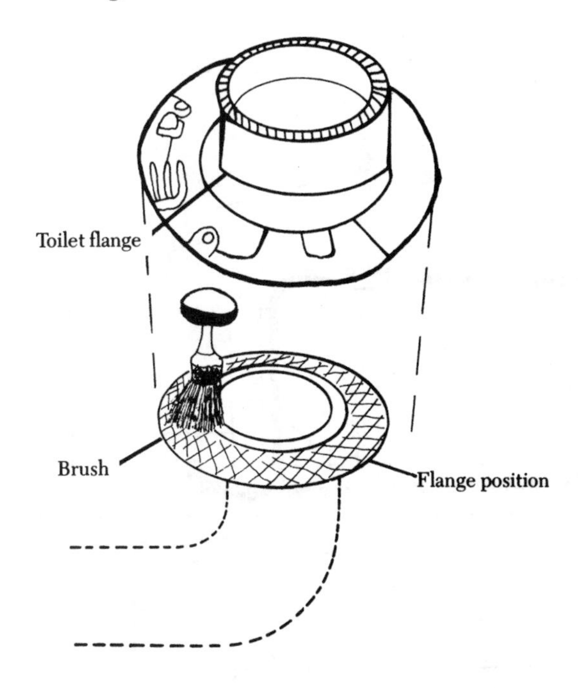

4. Position the SW&V fitting in the wall. In order to do this you will have to remove a little more than 3″ of the 2″ × 4″ wall plate and the subflooring beneath it. If you are working on the second floor and there is a wall underneath you, you will encounter a 2″ × 4″ top plate that must also be cut. Once you have cut away enough space for the fitting, position it.
5. Measure the distance from the elbow under the toilet flange to the 3″ outlet in the SW&V fitting.
6. Cut and cement a piece of 3″ PVC pipe in the flange elbow and the 3″ outlet in the SW&V fittings.
7. Drop a plumb line through the center of the SW&V fitting as far as it will go. This may be only as far as the firestop in the wall below, or it may go all the way to the first floor. If you are working from the first floor, the plumb should reach the basement floor.
8. Wherever the plumb line encounters an obstacle, use the point of the plumb as the center of a 3″ hole to be cut through the wall sole plate, floor, or whatever. Continue extending the plumb line downward and cutting through whatever obstacles it encounters until the plumb can hang from the SW&V fitting to the basement floor.
9. The plumb line tells you exactly where the soil stack will end and the drain line to the existing building drain begins. The stack may be as long as 15 or 20 feet if it is going up to a second story, or as short as a few inches if coming down from the ground floor. Determine the point where the stack and drain line will join, bearing in mind that the drain line must slope at least ¼″ for every foot of its run to the building drain. Most likely, you will want to set the stack-drain connection near the joists if the building drain is above ground in the basement. If the building drain is buried under the basement floor, your drain line may have to start closer to the floor and reside in a trench.

Installing the House Drain and Stack

1. When you have established where the soil stack will end, cut a piece of 3″ pipe, push it up through the way you have cut through the house, and cement it to the bottom part of the SW&V fitting.
2. Cement a 45° street elbow to the bottom of the stack. Be careful to aim the elbow exactly in the direction that will get your drain line to the building drain.
3. Install a 3″ Y to the end of the street elbow and insert a cleanout and plug in the open end of the Y, facing away from the house drain.

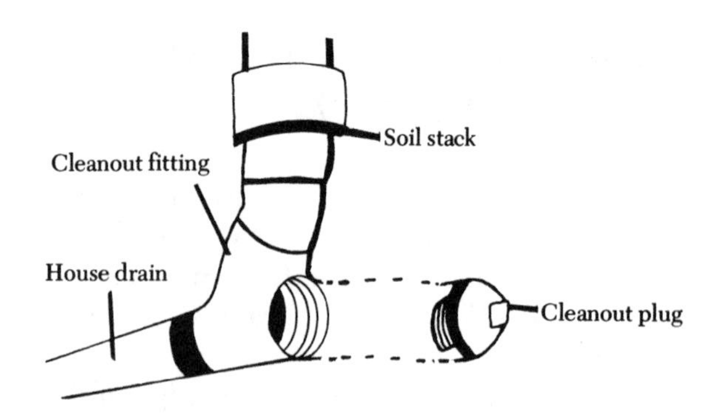

The house drain can be opened for cleaning by unthreading the cleanout plug.

4. The 3″ drain line that runs from the Y to the existing house drain should slope at least ¼″ for every foot of its run, and should have as few bends in it as possible. The bends

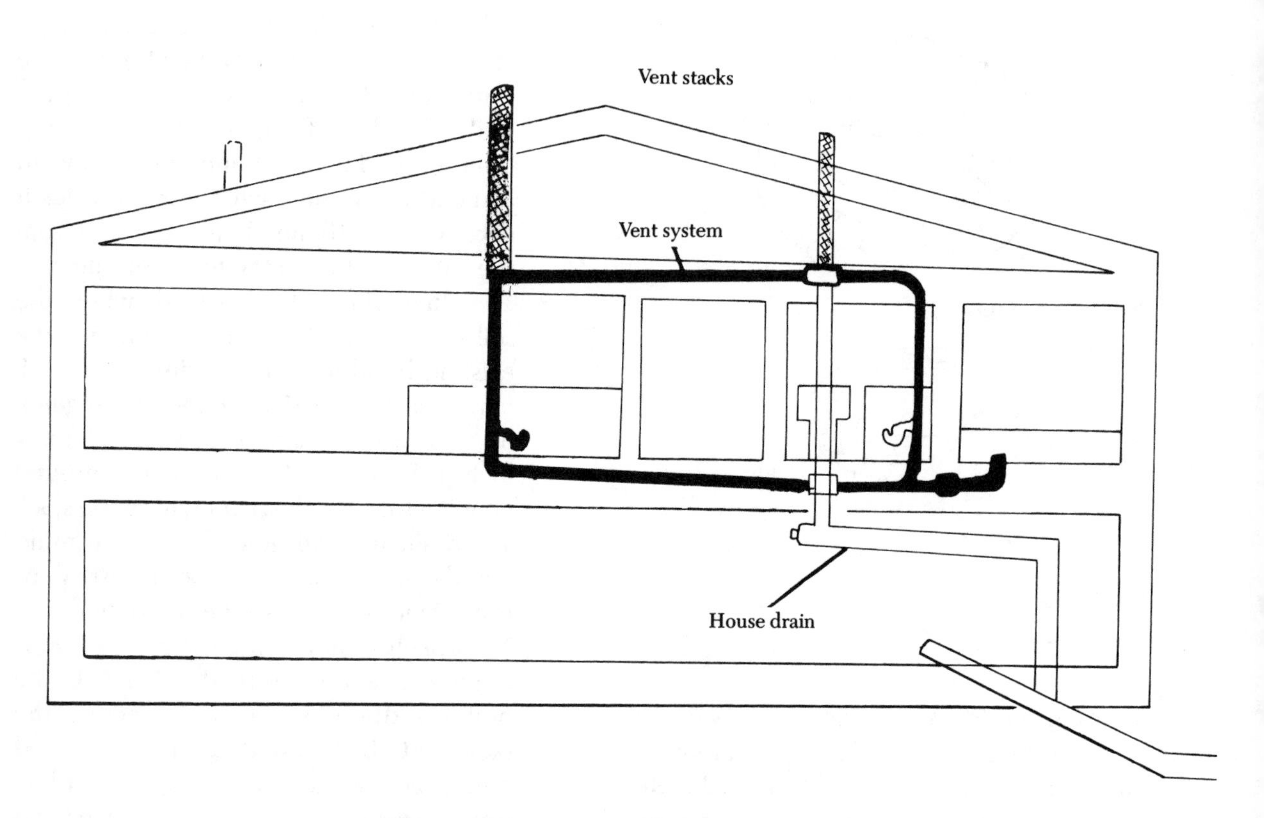

The vent lines in a typical DWV system

should not be 90° elbows; instead, use long-sweep 90° bends or assemble a pair of 45° bends. A straight 90° elbow makes too sharp a turn and is likely to become clogged. The simplest way of connecting to the existing house drain is at its cleanout opening at the base of the main stack. Remove the cleanout fitting and plug.

5. Install the proper adapter in the existing cleanout hole to accept your 3″ drain line, and cement a Y in the hub.
6. One branch of the Y is given a cleanout fitting and plug.
7. The other branch receives a length of 3″ pipe that connects with your drain line via a long-sweep 90° bend or a pair of 45° bends. You can assemble these parts, but do not cement them yet.
8. The bend at the top of the short pipe should face the Y hanging from the bottom of the soil stack. Cut and cement a length of 3″ pipe between the two fittings and support the pipe every 2 feet with pipe hangers attached to the joists.
9. Do not cement the bend(s) and short length of pipe to the Y in the building drain. Instead, put a temporary cap in the connecting branch of the Y. If you hook up your add-on DWV system to the existing house drain line now, sewer gases may escape into your house. The final hookup can be made after the fixture drain lines are installed and capped and the vent lines are in place, or, better still, after the fixtures have been installed.

Installing the Fixture Drains

1. Tubs and showers can use 1½″ waste pipes which are connected to one of the SW&V fitting's side ports. The openings are set at a 45° angle in the fitting, allowing you to angle the tub-shower drain down from a run under the floor to a 45° street elbow. You can also make the run along the wall by twisting the 45° elbow sideways and inserting the pipe through notches in the wall studs. Cement a 45° street elbow in one of the SW&V fitting openings and angle it in the direction you want the drain line to run.
2. The drain line should angle upward from the SW&V fitting to the tub or shower at least ½″ for every foot of its run. Cement the pipe to the street elbow.
3. The tub-shower end of the pipe terminates at a position that centers it at the drain end of the tub. Cement a 1½″ trap adapter to the end of the pipe that will accept the tub's P-trap. If you are installing only a shower, the drain line connected to the shower trap will be 2″ in diameter; you will need either a 2″ × 1½″ reducer coupling or an SW&V fitting that has 2″ drain ports, plus a length of 2″ drainpipe. Somewhere

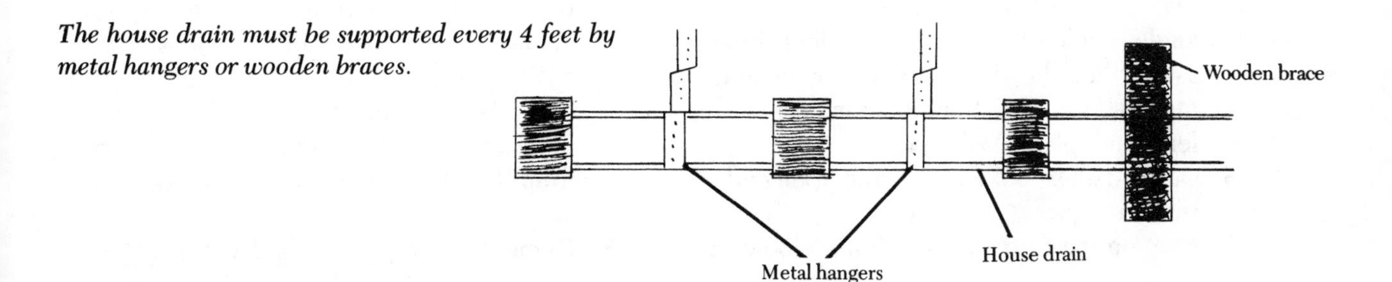

The house drain must be supported every 4 feet by metal hangers or wooden braces.

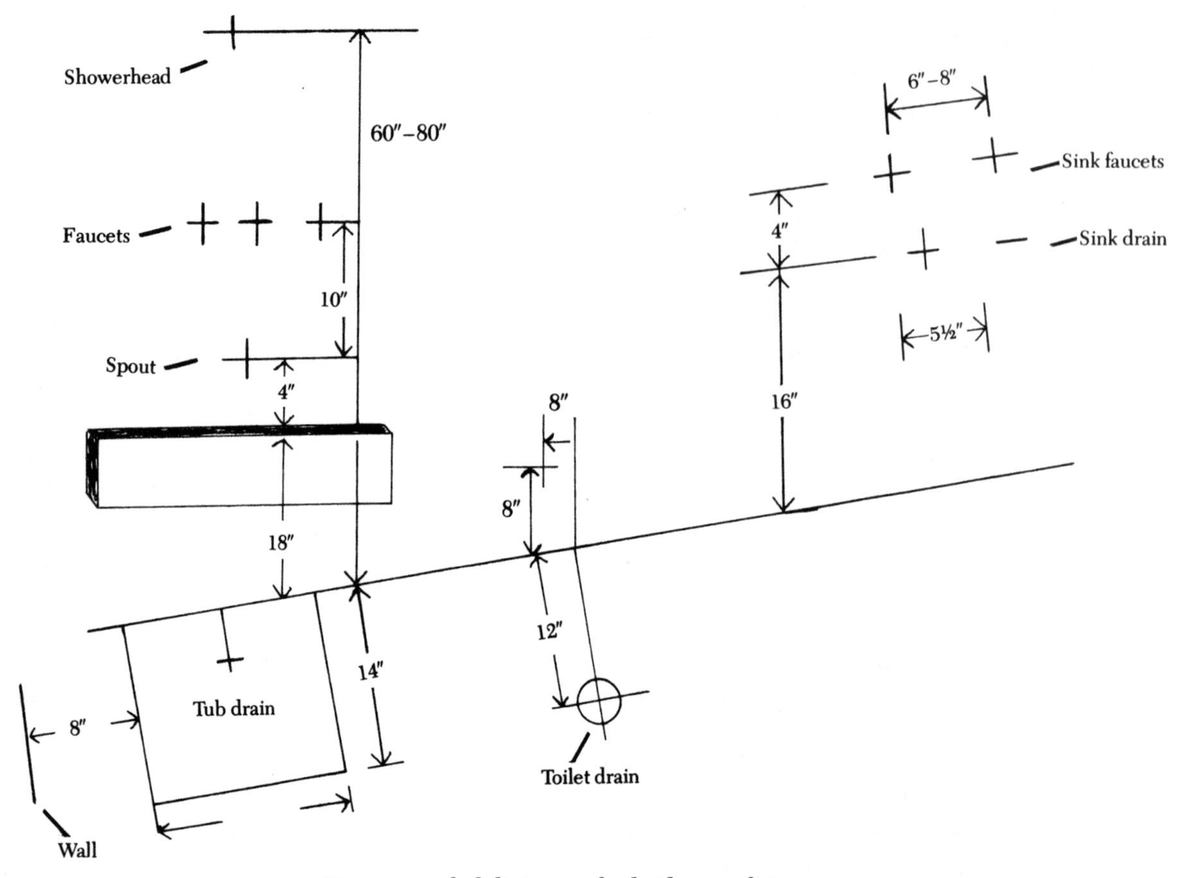

Recommended distances for bathroom fixtures

along the length of the run, insert a T that can be used to connect the vent line.

4. Sink drains usually go into the bathroom wall behind the sink, and are 16" to 18" above the floor. The drain line then travels along the wall (in notches cut in the studs) to connect with the soil stack via a 3" × 3" × 1½" T. It must slope toward the stack at an angle of at least ¼" per running foot. Begin assembling the sink drain line by cementing a 1½" × 1¼" trap adapter to a short-length nipple of 1½" pipe.
5. Cement a 1½" 90° elbow to the open end of the 1½" pipe.
6. Position the trap adapter/nipple/elbow 16" or 18" above the floor at the center of where the sink will reside, and measure the distance to the soil stack. Angle your ruler downward ¼" for every foot of the run and mark its position on the wall or a stud above the SW&V fitting.
7. Measure the distance from the top socket of the SW&V to your mark on the wall. Allow for the make-up distances in the SW&V socket, and for one socket of a 3" × 1½" reducing-T.
8. Measure and cut a length of 3" pipe to run from the SW&V to the 3" × 1½" reducing-T.
9. Cement the 3" pipe in the SW&V fitting.

10. Cement the 3″ × 1½″ reducing-T to the 3″ length of pipe. Make certain the T is facing the direction of the sink drain line.
11. Measure and cut a length of 1½″ drainpipe to run from the reducing-T in the stack to the 90″ elbow at the sink drain position.
12. Assemble the 1½″ pipe, between the reducing-T and the elbow to make certain they are correct, and that the slope of the drain is proper. Hold the pipe against the edges of the studs and mark the top and bottom of the pipe on each stud for notching.
13. Remove the pipe assembly and notch the studs.
14. At some point during the run, insert a T to accept the vent pipe.

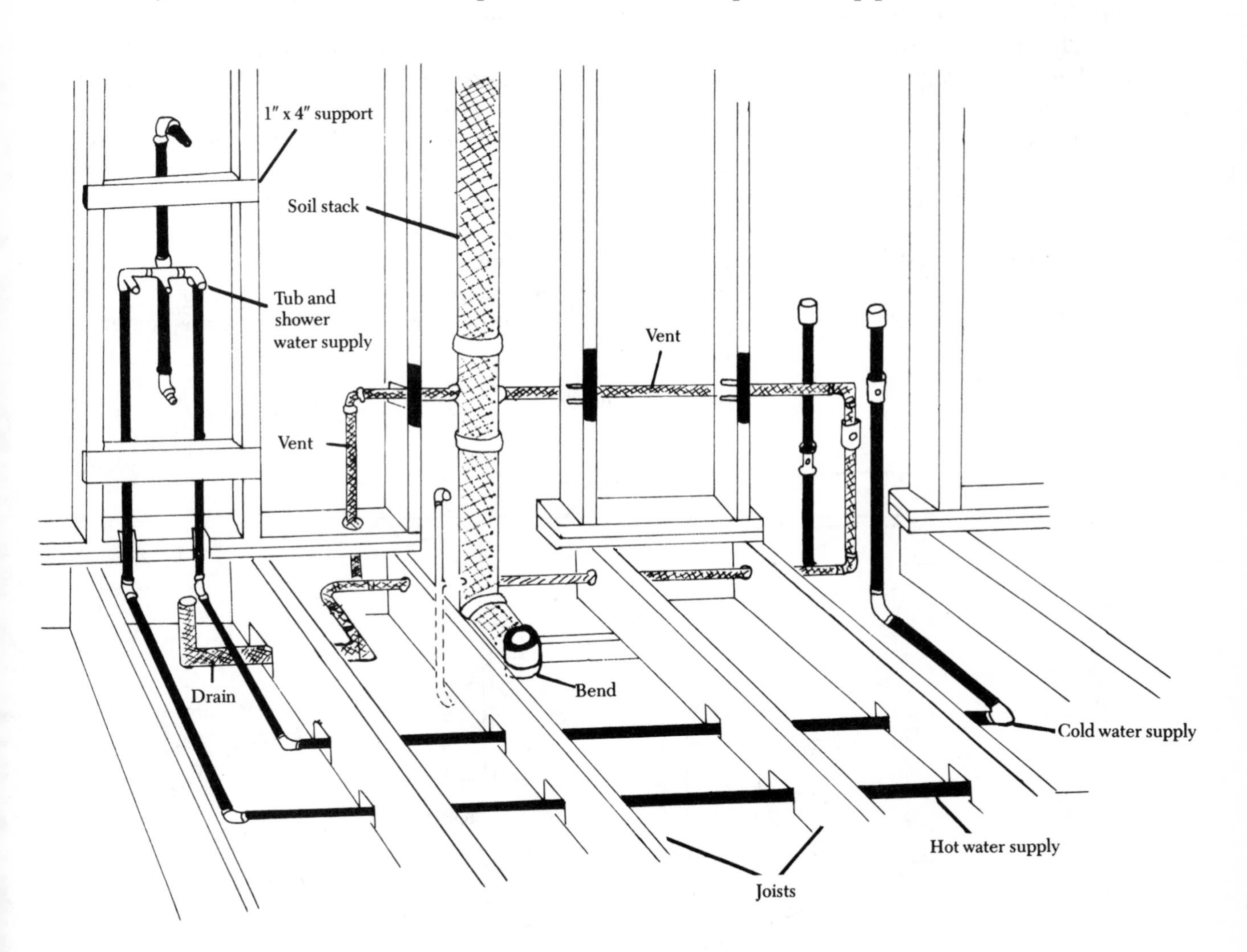

The DWV and water supply systems in a three-fixture bathroom

15. Cement the pipe to the reducing-T and the elbow and insert the drain line in the notches. Place metal plates over the opening in each notch.

Assembling the Vent Lines

All of the pipe attached to the soil stack above the highest drain connection is considered vent pipe. It can be 2″ in diameter, but that becomes another piece of pipe diameter you have to deal with, and you'll need a reducer coupling. It is easier to continue the stack up to the roof using 3″ pipe. All of the fixtures must be vented straight up to the roof or revented with a 1½″ pipe that connects to the soil stack at a point above the highest drain connection.

1. Continue the soil stack up to the roof from the top of the bathroom sink connection. If you are working on the ground floor of a two-story house, you will have to cut a way through the top plate of the wall and carry the pipe up through the wall above it. Continue cutting a way through the top and sole plates of the wall until you reach the attic.
2. You can insert 3″ × 3″ × 1½″ T's in the stack pipe at whatever point you wish to connect the vent lines from the sink and tub-shower. If you are going up past a second story it is usually easiest to bring the vent lines up to the attic and make their connections there, but they can join the stack at any point above the sink drain line. Of course, if you bring the vents up to the attic, you must cut ways for them through the walls that carry them.
3. The vent lines from the sink and tub-shower are 1½″ pipes that are cemented

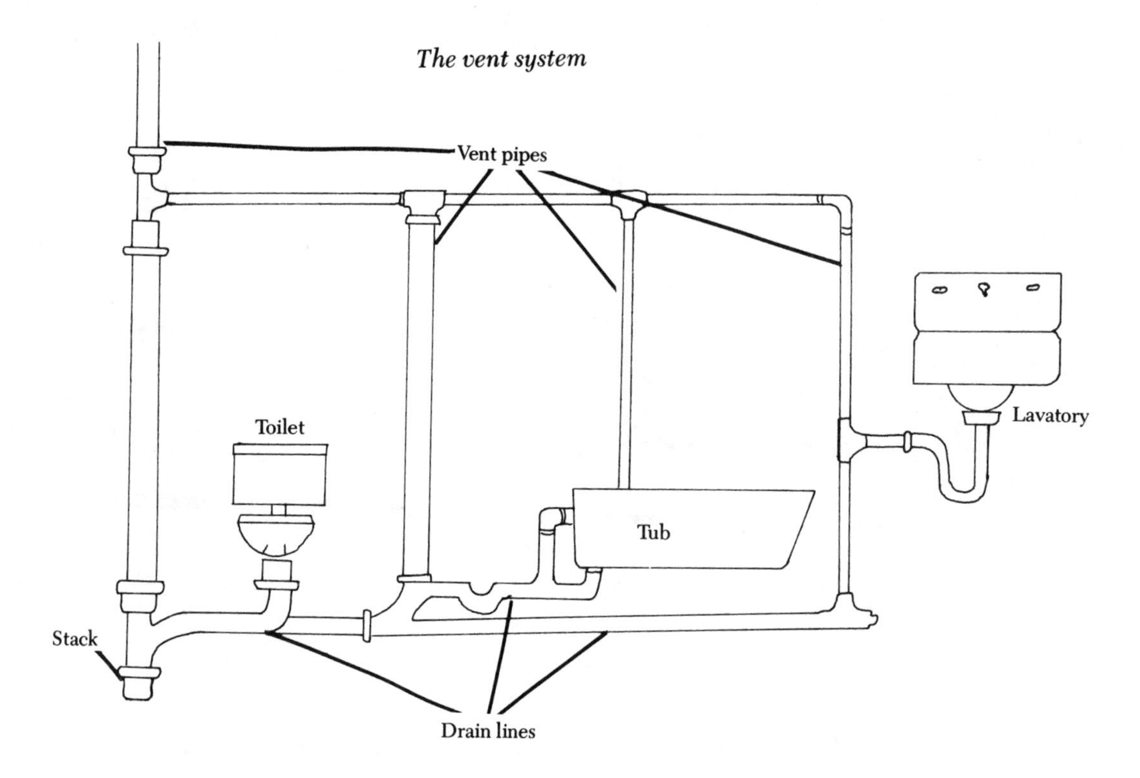

The vent system

into the vent T's inserted in the drain lines, and then brought up to whatever point they connect with the stack.

4. When the vent lines are opposite their connecting T's in the stack, cap them with a 90° elbow.
5. Cement a length of pipe between the elbow and the T in the stack.
6. Continue the 3″ vent line up to the roof and mark the underside of the roof sheathing where the pipe will go through the roof.
7. Drill a hole through the roof as a starter for your saber saw, and also to mark where the hole will be.
8. Go up on the roof and locate the hole. Remove the shingles surrounding it.
9. Cut out the hole for the stack with your saber saw.
10. Extend the stack up through the hole and cut the pipe off at least 12″ above the shingles.
11. Install the vent cap around the base of the vent pipe. The base of the cap should be liberally coated with roofing cement and nailed to the sheathing. The shingles are then replaced over the flashing.

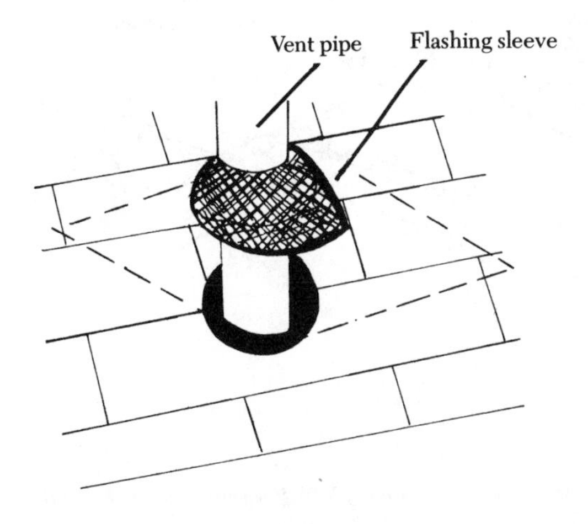

Vent caps are made to fit snugly around the base of a pipe and form a watertight seal.

12. The DWV is now complete, except for its final hookup with the building drain. If you have not used all of the ports in the SW&V fitting, be sure to install a cap in the unused port. Also, check all of your horizontal pipe runs to be sure they are supported every 32″—that is, at every other stud. If a horizontal pipe is not held securely in notches, you can wrap a piece of perforated steel strap (plumber's tape) around the pipe and nail the ends of the strap to the side of a joist or stud.

Testing the DWV

If you have taken out a building permit, your local building department may require you to pressure-test your newly built DWV system before a department official. The time to make the test is before you close up the walls, but you can make it either before or after you assemble your water supply system.

Prior to making the test, complete cementing your add-on system to the existing house drain. Then rent a set of rubber plugs from a plumbing supplier to seal off the closet bend (the toilet flange) and the house drain cleanout plug. You also have to cap the sink and tub-shower drain lines. Wait at least 16 hours after the last joint is cemented for the solvent-weld to become completely hard, then follow this procedure:

1. Hook up a garden hose and take it to the roof.
2. Poke the nozzle of the hose down the soil stack and turn the water on.
3. When water appears at the top of the stack, shut off the hose.
4. Wait 30 minutes and then inspect each of the joints you have made for any sign of leakage. If there are any leaks, repair them.
5. When you are satisfied that none of your

joints is leaking, call the building inspector.
6. When the inspector has examined your work, remove the plugs in the house drain and toilet flange and allow the water to drain out of the system. Then replace the cleanout plug in the house drain.

Plumbing the Water Supply System

You can start the construction of your water supply lines from any of the fixtures, or from the point where you intend to branch off from your existing hot and cold water mains to the add-on bathroom. Your water supply system can, of course, be the same pipe that is used throughout the rest of your house, or you can use CPVC rigid pipe and PB tubing, beginning at the connections to the existing mains.

CONNECTING PLASTIC PIPING TO COPPER

1. Close the nearest valve controlling each of the water mains.
2. Open the nearest faucets to drain the pipes of water.
3. Using a tubing cutter or hacksaw, cut out a section of the copper tubing that is long enough to allow you to insert a CPVC-T and its adapters. Copper tubing is somewhat flexible, and often you need only make a single cut through it and then push the ends of the pipe apart enough to insert the T. If moving the pipes apart is not

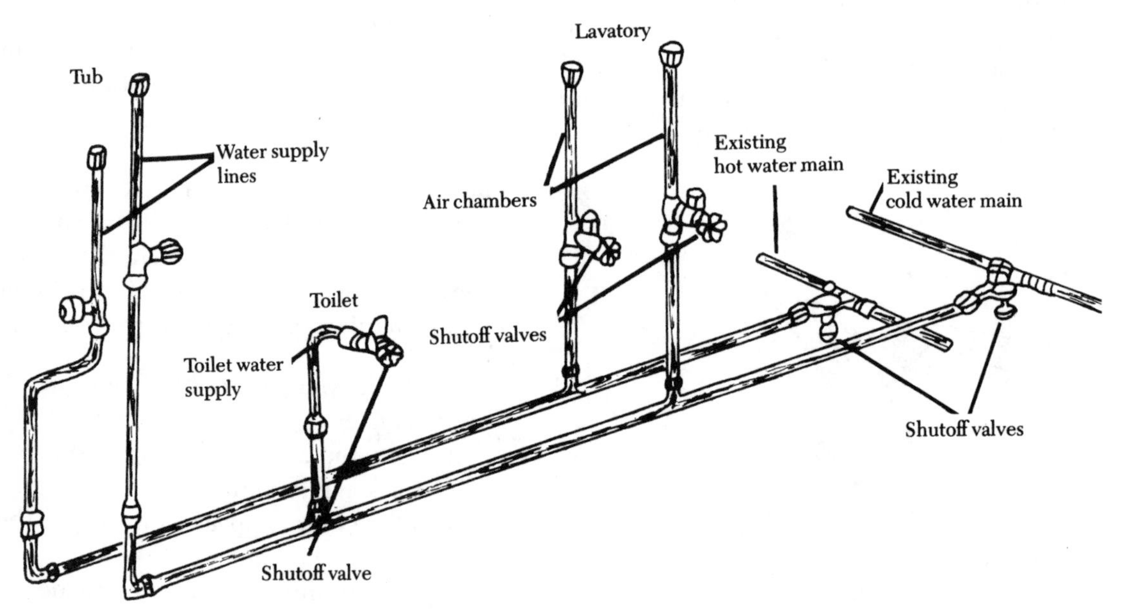

The CPVC-PB water supply system

feasible, you will have to make two cuts in the pipe.

4. Clean the ends of the cut pipe to remove all burrs.

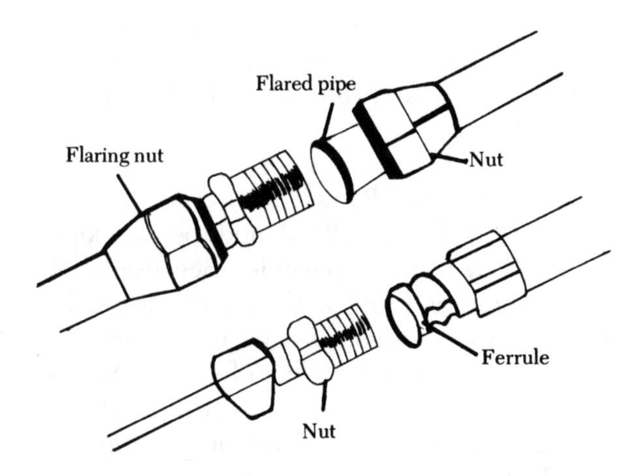

A flared coupling (top) and a compression coupling (bottom)

5. You can adapt plastic to copper tubing by using either a flared or a compression coupling, or a special pair of adapters that are then solvent-welded to the ends of a CPVC-T. If you use a flared coupling, both the copper tubing and the CPVC plastic must be flared, using a standard flaring tool. The compression coupling requires brass ferrules fitted over the pipe and then squeezed into a watertight seal by tightening the compression nuts. The T-fitting with special adapters is made of CPVC and can be connected to copper pipe by hand-tightening the adapters around the copper. If you choose to branch out of the T with PB tubing, you must solvent-weld a coupling to the T to accept the PB. Whichever method you choose, connect the T so that it is positioned toward the direction in which the pipe will travel.

6. When a T has been inserted in each water supply main, continue your pipe runs by connecting either CPVC or PB piping to the open port in the T.

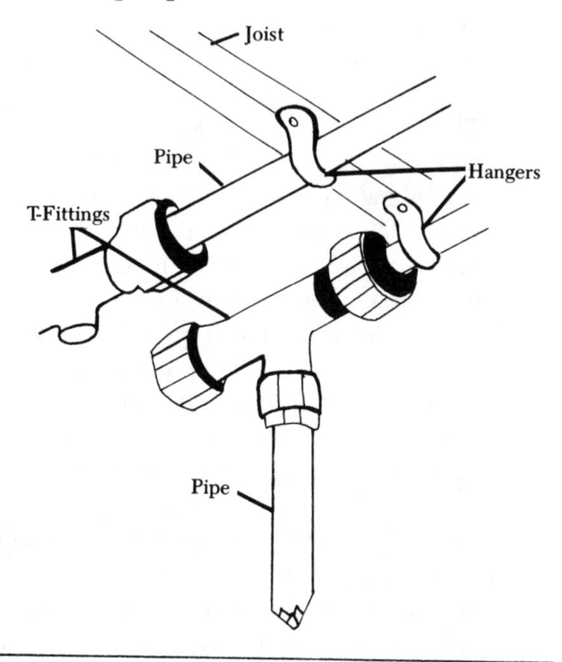

A specially designed CPVC adapter is connected by hand-tightening its nuts.

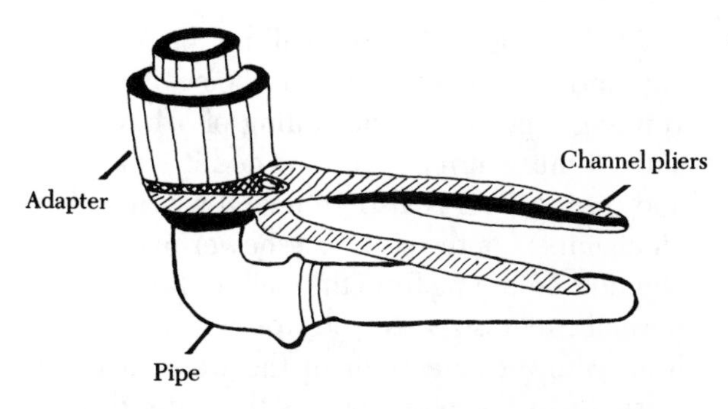

CONNECTING PLASTIC PIPE TO THREADED PIPE

1. Cut the brass or galvanized steel with a hacksaw, removing enough of the pipe to allow you to insert a transition union.

2. Clean the cut ends of the pipe of all burrs.
3. Use a stock and die to thread the ends of the pipe.
4. The transition union has two halves that clamp together with a threaded metal side, which is connected to the metal pipe, and a solvent-welding side to accept CPVC pipe. The two halves are held together with a hand-nut, but there is a rubber gasket between halves that allows the metal and plastic to expand and contract at their own rates without causing a leak. Thread a union to each end of the severed pipe.
5. Cement a CPVC transition union between the connections, and begin your pipe run from its open port. If you want the run to start with PB, you will have to add a PB adapter to the CPVC transition coupling. Alternatively, you can adapt to galvanized steel or brass with a compression-T that fits over the pipe and is held to it with compression nuts. Assemble a single transition union to the T-port to complete your connection to the plastic piping.

Running the Water Supply Lines

CPVC is rigid pipe and will not bend around any corners. It is the usual choice if you are running pipe across the ceiling of a basement, since it must turn 90° corners and it looks clean and neat. CPVC is also preferred when making air chambers or the riser to a showerhead, and as stubouts emerging from the walls or floor. Under normal circumstances, you would use ¾″ CPVC from your T-connections in the water mains at least up to the point where the runs turn up through the walls of the house, where your pipe run can change to PB.

The flexibility of PB allows it to be snaked through walls and around some relatively sharp corners with considerable ease. You can bend PB tightly—up to a 6″ radius with ¾″ piping, 4″ with ½″ piping, and 2″ with ⅜″ piping—without having to cut it and insert a fitting. Thus, from wherever the hot and cold mains enter the walls, you would use ¾″ PB and snake it between the studs as far as your add-on bathroom. Then, as the mains approach your fixtures, insert a reducer coupling and connect to the faucets with ½″ piping.

Supporting Plastic Pipes

Both CPVC and PB should be supported every 32″ whenever possible. Because of the effect of heat and cold on the plastic pipe and its fittings, your system must be free to move as much as ½″ for every 10 feet of its run. You can give it this freedom by holding it in place with sliding hangers that wrap around the pipe and are then nailed to a framing member with a single nail. The hangers should be placed on every other joint or stud—that is, every 32″.

Both pipes can, of course, reside in notches cut in the studs. PB, because it is so flexible, can be threaded through holes drilled through framing members.

Making the Pipe Run

Beginning from your adapter fitting inserted in the hot and cold water mains, carry both lines as directly as possible to your add-on bathroom. If the tap into the mains is within a few feet of the bathroom, you can make the entire run with ½″

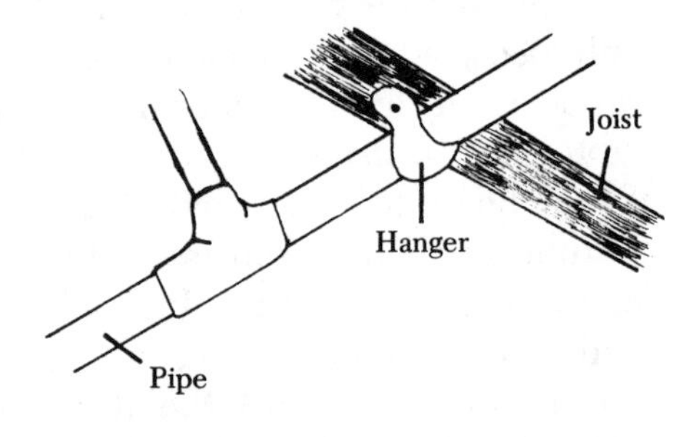

Support water supply pipes every 32″.

pipe; otherwise, use ¾″ pipe until you get to the bathroom walls and then branch off to each fixture with a reducer coupling and ½″ pipe.

CONNECTING THE SINK

1. Bring the ½″ hot water line up to the point under the sink where it will emerge from the wall.
2. Install a ½″ CPVC-T on the end of the piping with the T-port facing out of the wall. Secure the pipe with a hanger near the fitting.
3. Cement a 12″ length of CPVC in the top port of the T.
4. Cement a cap to the top of the 12″ length of pipe to complete the air chamber.
5. Cement a short length of CPVC to the T-port. The stubout pipe should be long enough to bring the water through the wall and hold a shutoff valve.
6. Connect the fixture shutoff valve to the stubout.
7. Repeat steps 1—6 for the cold water supply line.

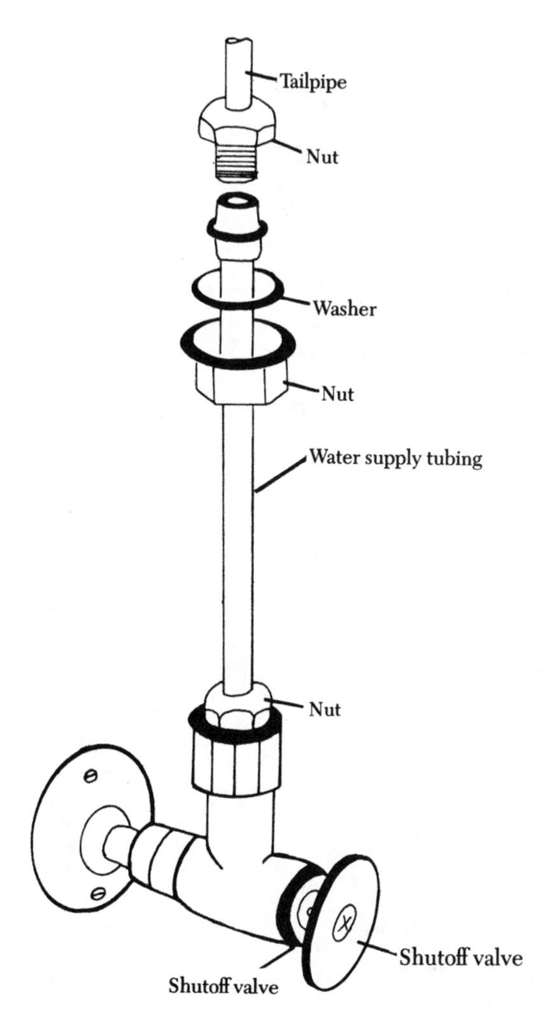

CONNECTING THE TOILET

1. Only the cold water supply line is tapped to supply the toilet. Insert a T in the line and run ½″ PB as far as the position for the shutoff valve, under the left (as you face the wall) side of the toilet.
2. Cement a CPVC wing elbow on the end of the PB pipe and secure the wing to a framing member.
3. Cement a short stubout to the open port of the elbow.
4. Cement a fixture shutoff valve to the stubout.

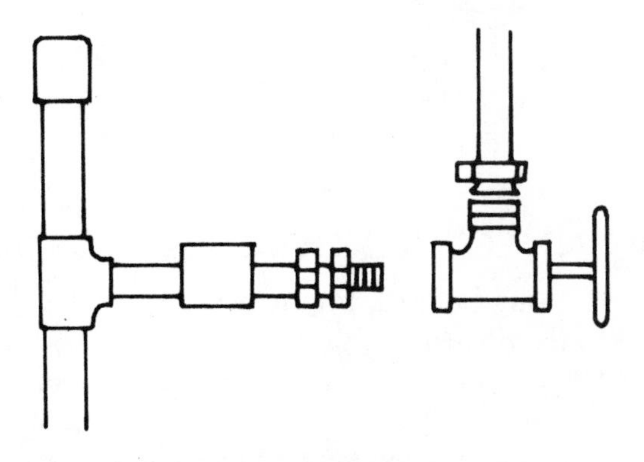

Water supply shutoff valve assembly

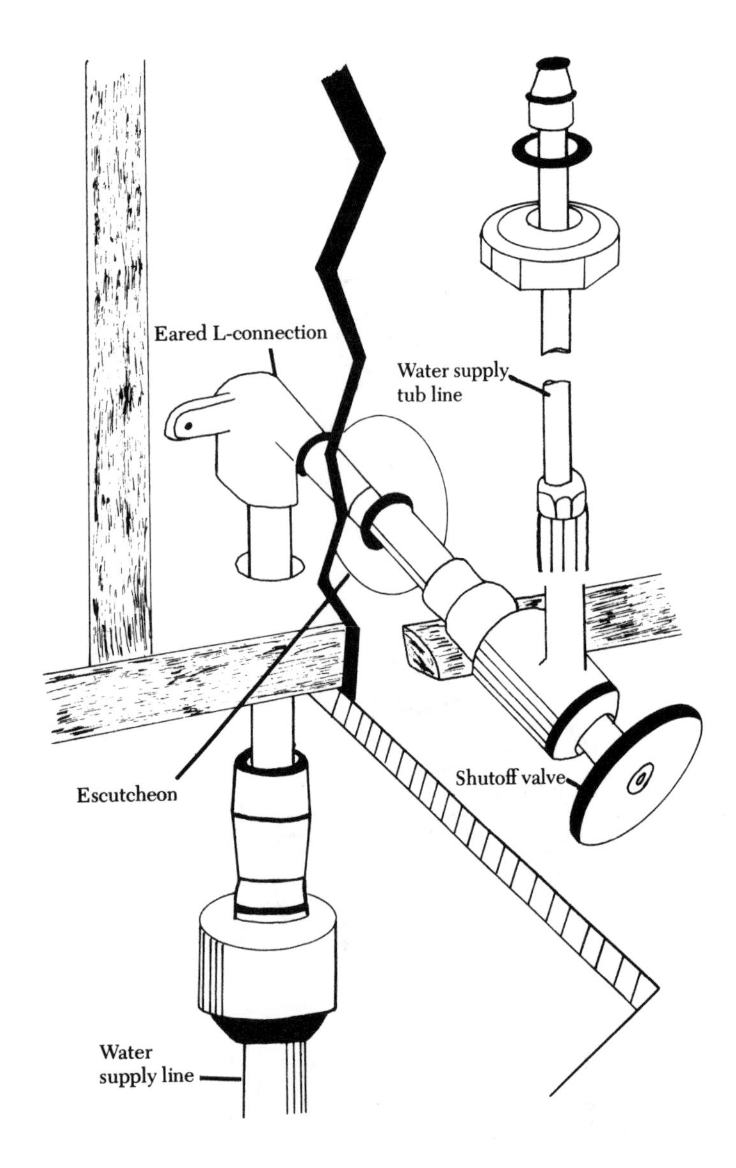

CONNECTING THE TUB-SHOWER

1. Bring both supply lines up to the level of the diverter valve.
2. Cement ½″ T's to the ends of the supply pipes.
3. Cement ½″ × 12″ CPVC pipes to the tops of the T's. Cap both pipes to complete the air chambers.
4. Cement short ½″ nipples from the T's to transition unions. The unions are necessary to complete the coupling with the diverter valve.
5. Connect the unions to the hot and cold sides of the diverter valve. Secure the diverter to a framing member in the wall.
6. Use an adapter to form the connection between the top of the diverter valve and the CPVC riser pipe to the shower.
7. Cement a ½″ CPVC wing elbow to the top of the riser and secure it to the framing member.
8. Connect an adapter to the bottom port in the diverter valve.
9. Cement a short length of ½″ CPVC to the adapter.
10. Cement a 90° elbow to the bottom of the pipe.
11. Cement a stubout pipe to the open end of the elbow.
12. Cement a threaded adapter to the stubout to accept the tub spout.

Testing the Water Supply System

Make certain that all of the shutoff valves you have installed are in their closed positions. In the case of the tub-shower, be certain the diverter valve is in its *off* position, and that caps are threaded to the tub spout and the shower riser pipe. Wait at least 16 hours after the last joint is cemented before pressure-testing the system.

Now turn on the shutoff valves controlling the water supply.

Carefully inspect each of the joints you have made—that is, look at every one of the fittings in your system. If water is visible at any joint, cut the fitting out and replace it.

When you are certain there is no leakage any-

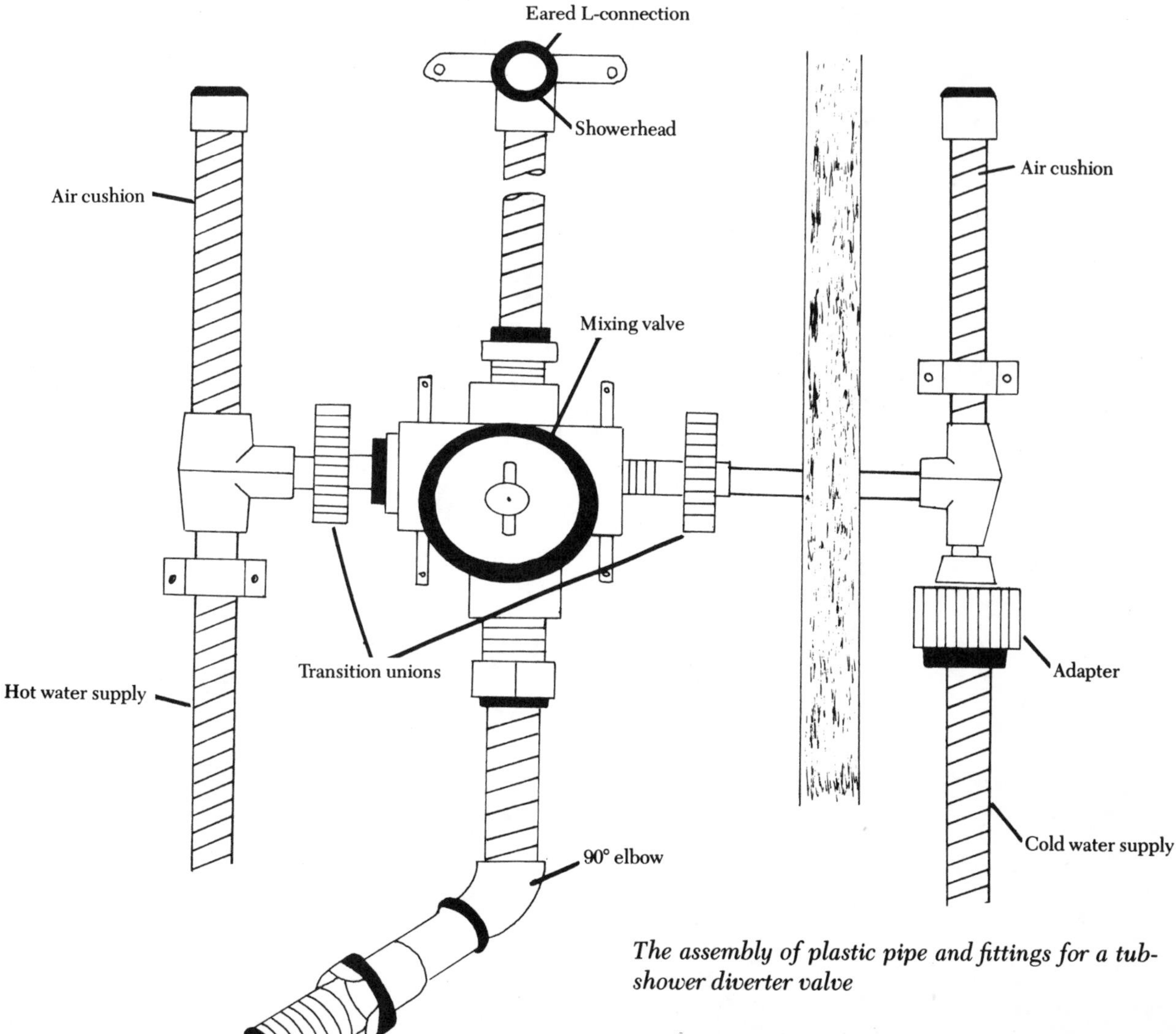

The assembly of plastic pipe and fittings for a tub-shower diverter valve

where in your plumbing, you can close up the walls of your add-on bathroom.

You may find it is easier to install the bathtub before the walls and floor are completed. In fact, if you are tiling around the bathtub, you have to put the tub in place after the sheetrock is nailed to the studs but before the tile is cemented in place, since the positioning of the tiles depends on the top edge of the tub. Otherwise, the wall behind the sink and toilet should be completely finished before either fixture is installed, and at least as much of the floor as will be directly under those fixtures must also be completed—which suggests that you might as well lay the whole floor and get it over with.

Installing a toilet is described on pages 48–54. The procedure for installing a sink is found on pages 55–60. Tub installation is discussion on pages 68–71. How to install a shower stall is covered on pages 61–67.

APPENDIX

Tools of the Trade

Adjustable wrench—The 12″ size will open to more than an inch, which is enough to grip most faucet bonnet nuts.

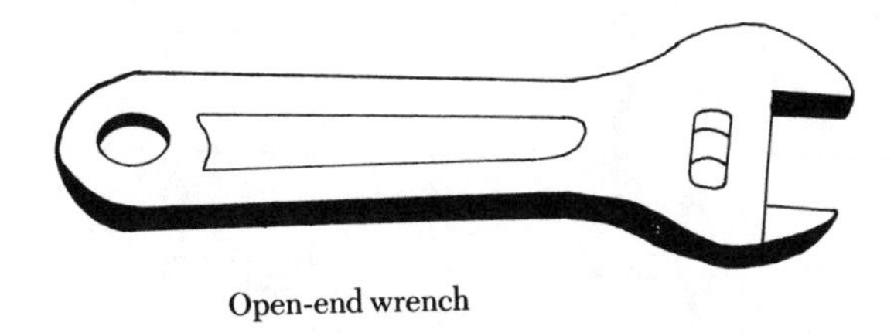

Open-end wrench

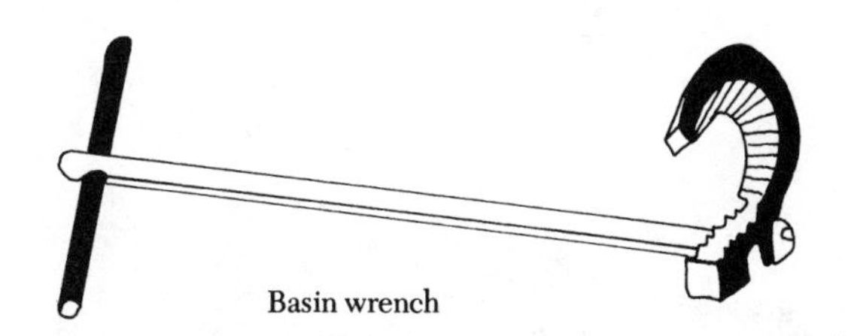

Basin wrench

Basin wrench—This is the only tool that allows you to reach behind a sink and undo the faucet nuts.

Pipe wrench—Every plumbing kit should include a 10″ and a 12″ pipe wrench. With both, you can handle any repair in the home.

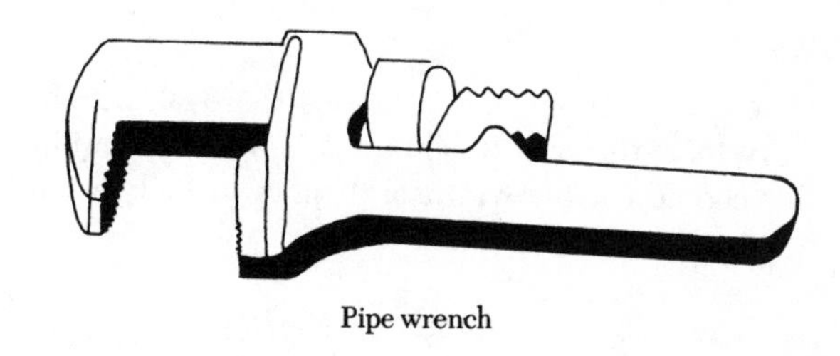

Pipe wrench

Chain and strap wrenches—You only need one of these, if any. Each will allow you to grip large-diameter nuts or pipes.

Wrenches—You can get these in all sizes to handle both large and small nuts and bolts. A socket wrench is good for getting at faucet parts, and the ratchet socket wrench allows you to work in confined areas.

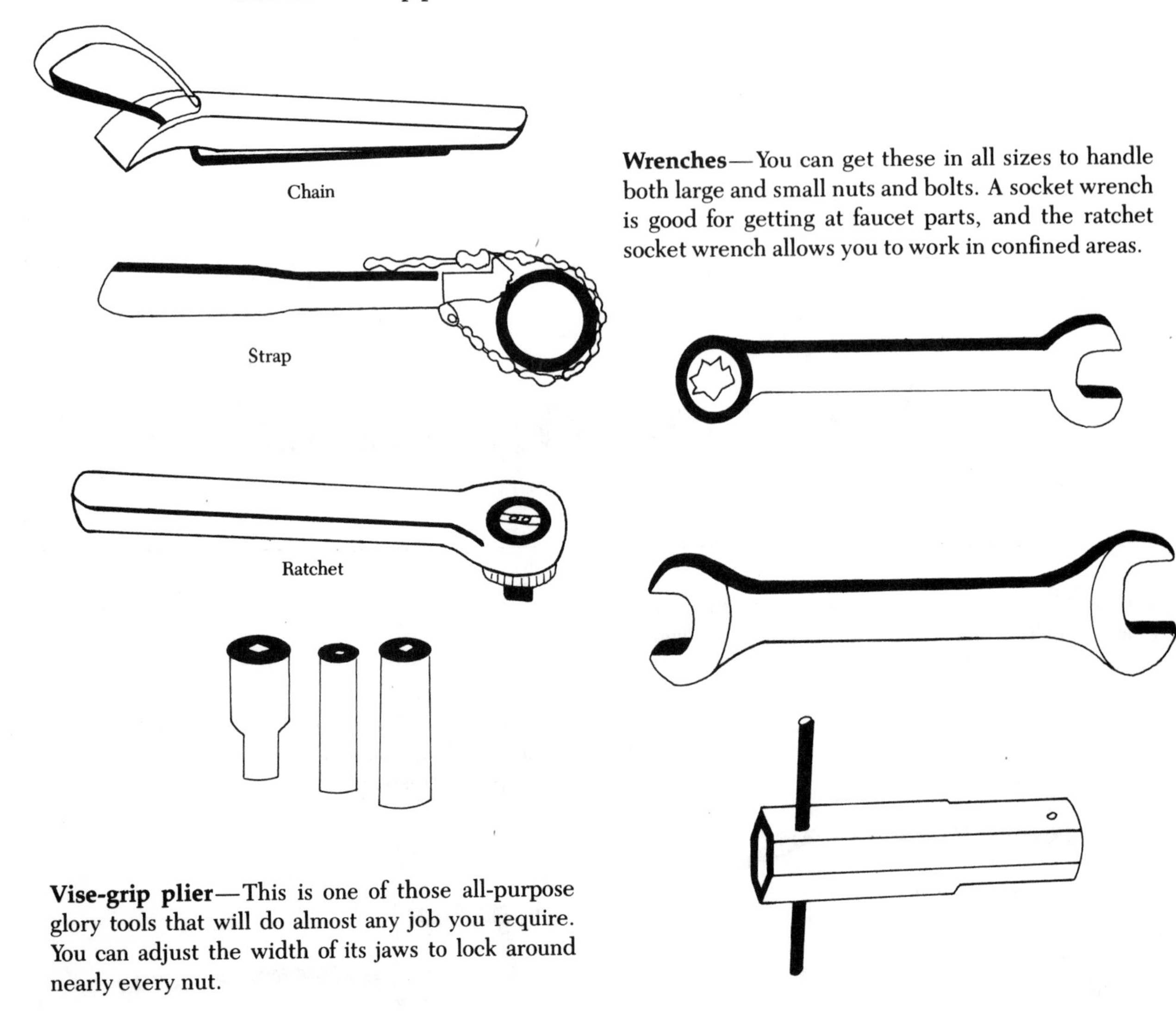

Vise-grip plier—This is one of those all-purpose glory tools that will do almost any job you require. You can adjust the width of its jaws to lock around nearly every nut.

Vise-grip pliers

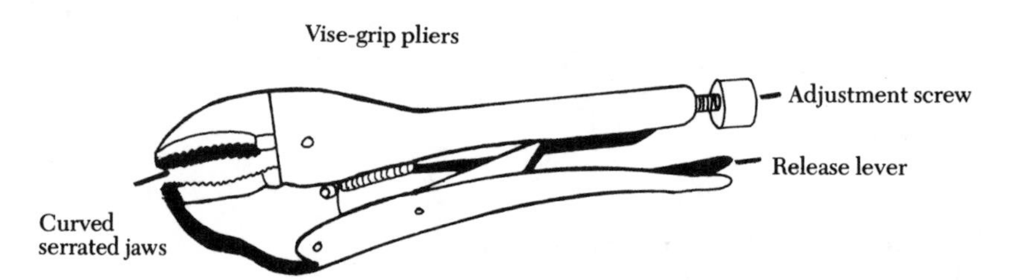

Pliers—Both the slip-joint and long-nosed pliers are good for gripping small fittings. The channel-type pliers are designed to get at larger nuts and offer tremendous leverage.

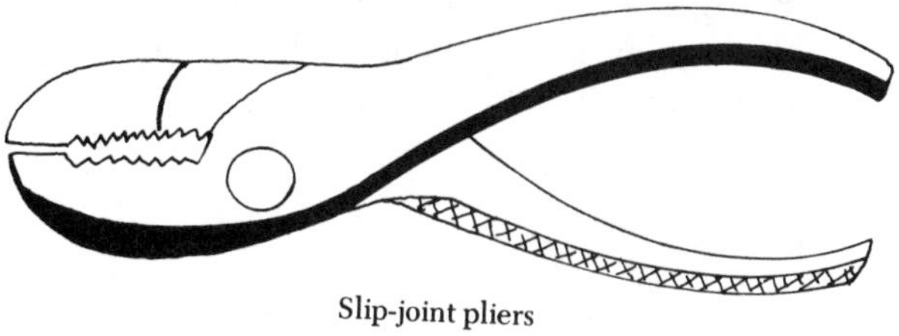

Slip-joint pliers

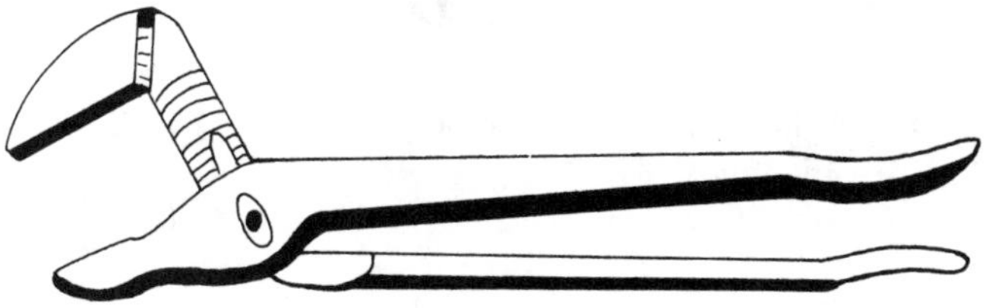

Channel-type pliers

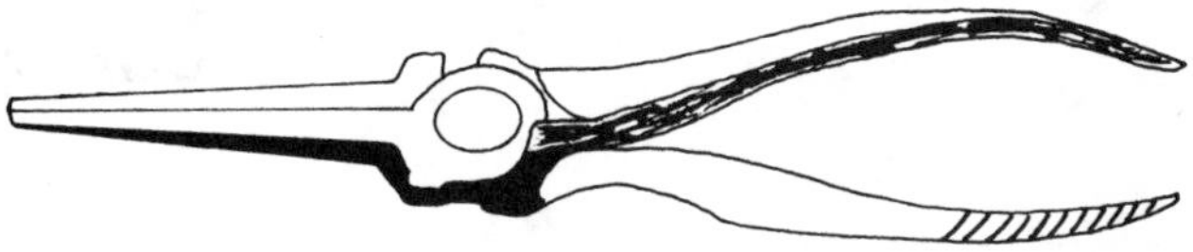

Long-nosed pliers

Screwdrivers—You really should have three or four standard blade widths as well as a phillips-head screwdriver for getting at the set screws in faucet handles. The stubby, offset, and ratchet screwdrivers are handy for tight situations.

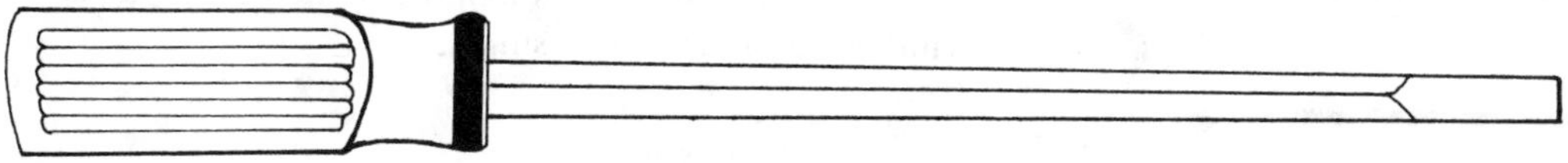

Standard

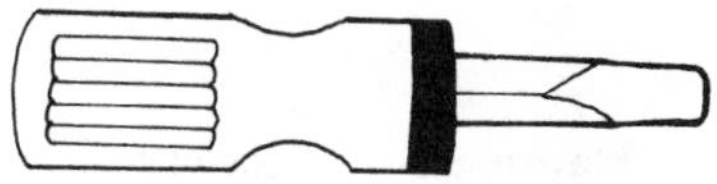

Stubby

Phillips-head

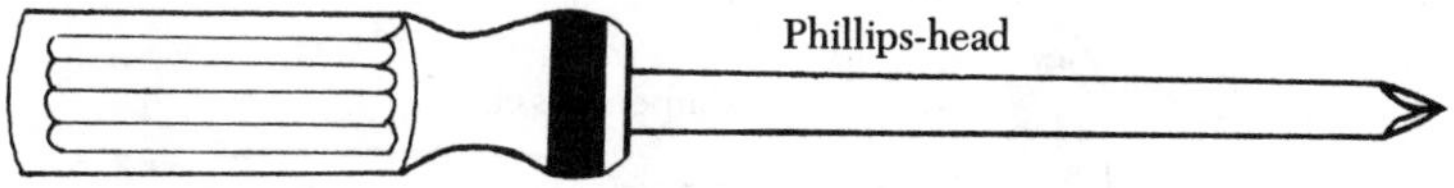

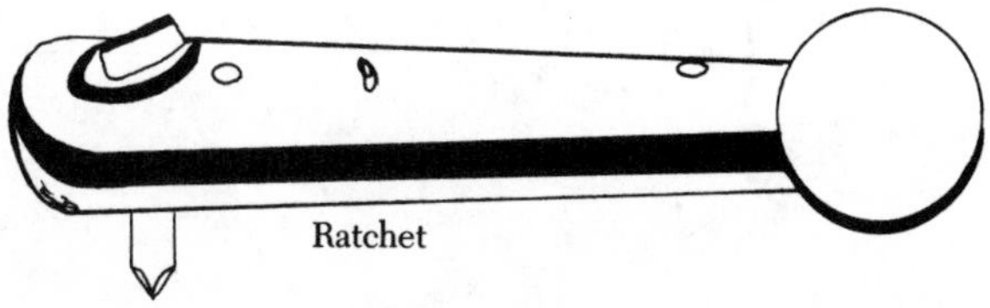

Standard

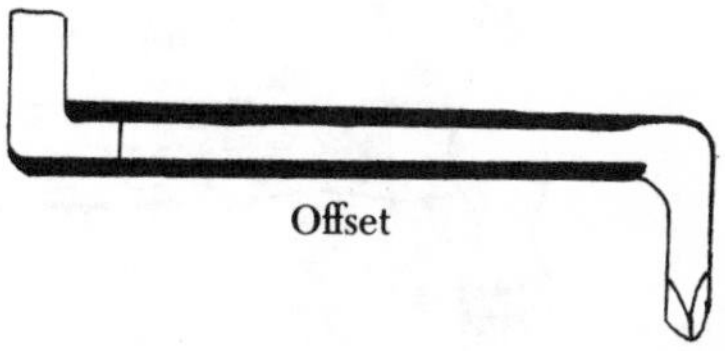

Offset

Ratchet

Phillips-head

Plungers—Both types shown are effective. It sometimes helps to gain a tight seal by coating the bottom of the cup with petroleum jelly.

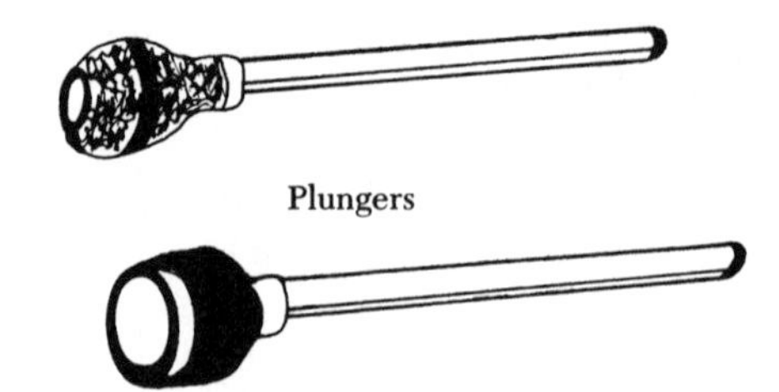

Augers—You can buy either a spring metal tape or a coiled spring. Both are effective.

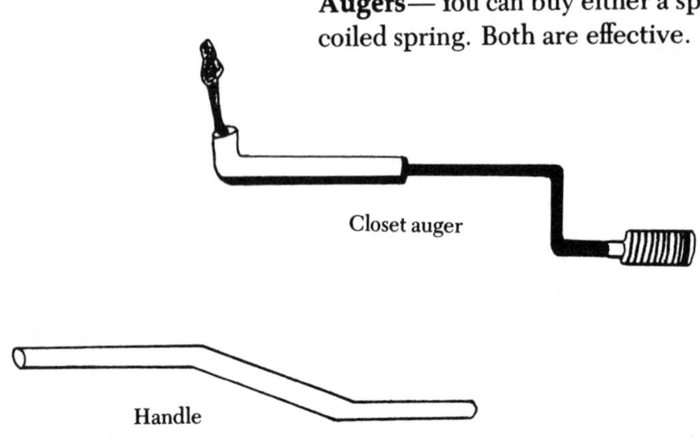

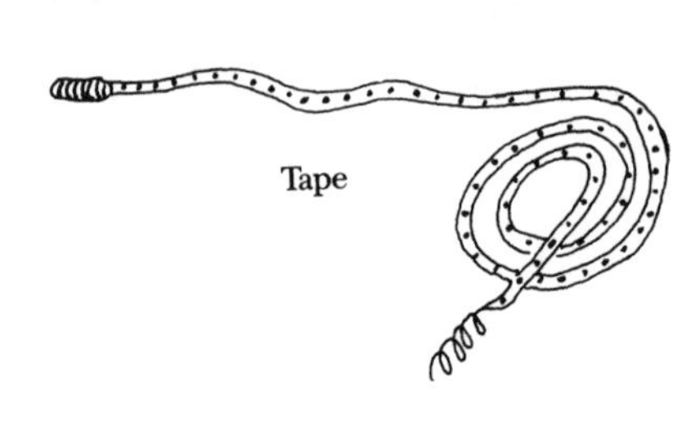

Spud wrench—There are both fixed and adjustable versions of the spud wrench. They are used to work with the large spud nuts found between toilet tanks and bowls, and in some sinks.

Hammers—These vary from 2 to 20 pounds, and are good for heavy-duty work. The standard carpenter's claw hammer is always useful; ball-peen hammers are good for shaping metals.

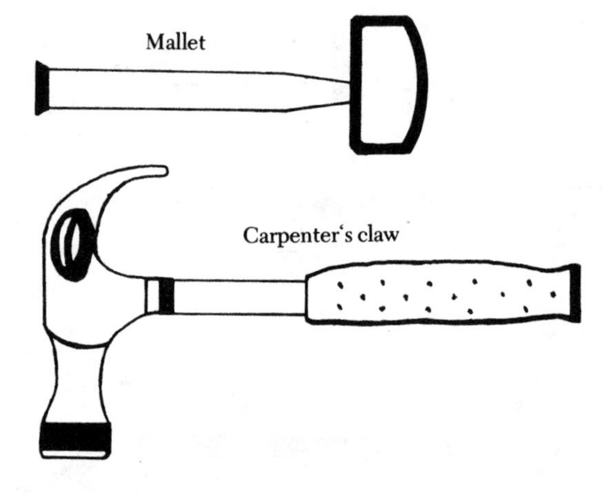

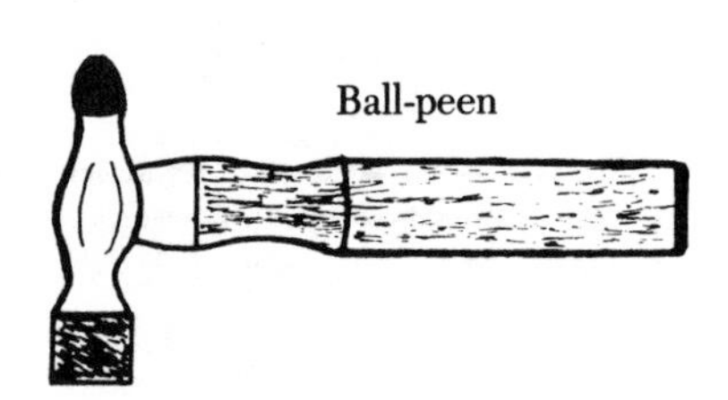

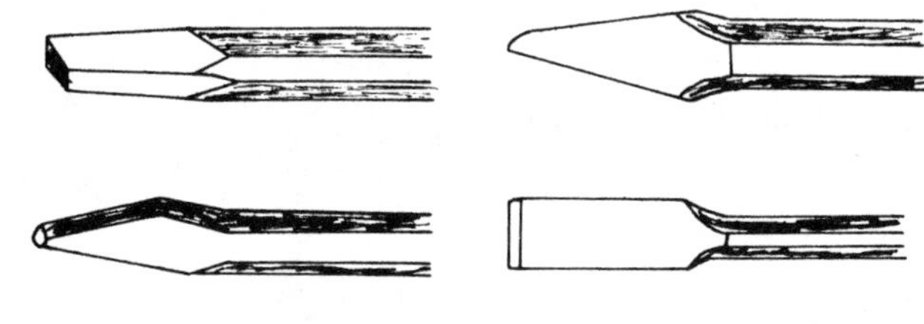

Cold chisels

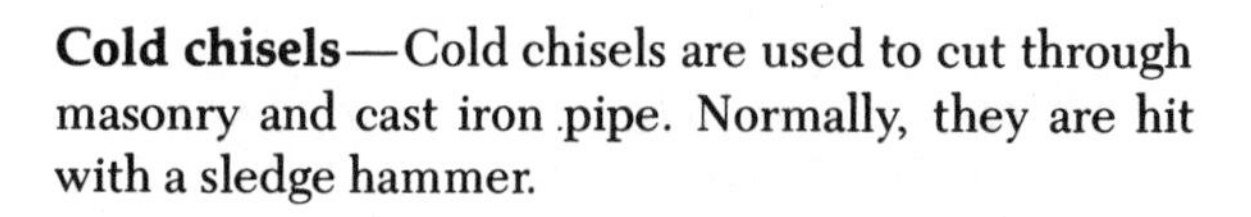

Cold chisels—Cold chisels are used to cut through masonry and cast iron pipe. Normally, they are hit with a sledge hammer.

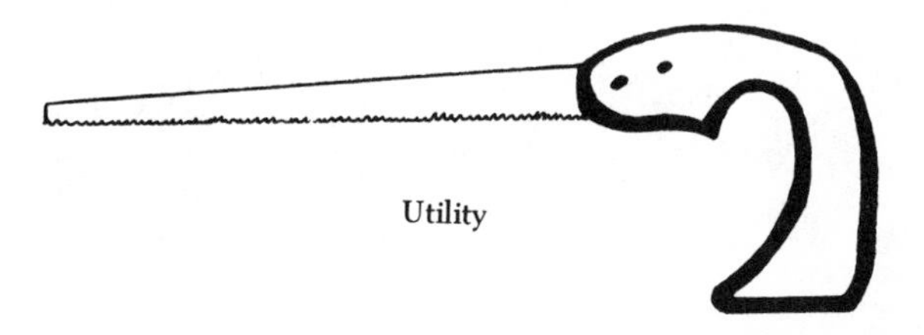

Utility

Saws—The utility saw is another one of those magic tools that will accomplish an amazing amount of work in limited spaces. The most versatile handsaw for plumbing work has a 10″ to 12″ blade with 24 to 32 teeth per inch. The hacksaw will cut cooper, brass, galvanized steel, cast iron, and plastic pipe.

Ripsaw

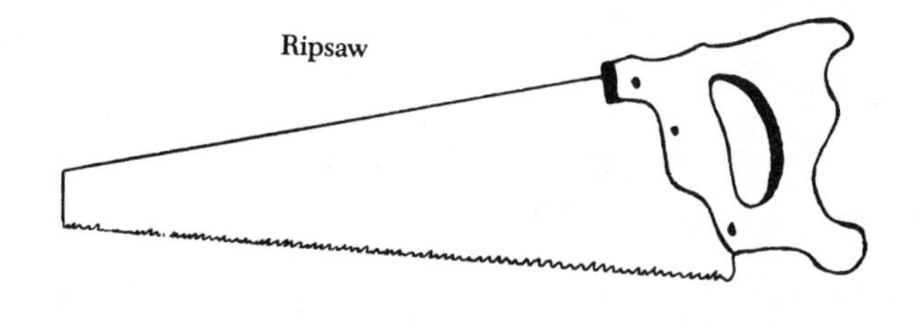

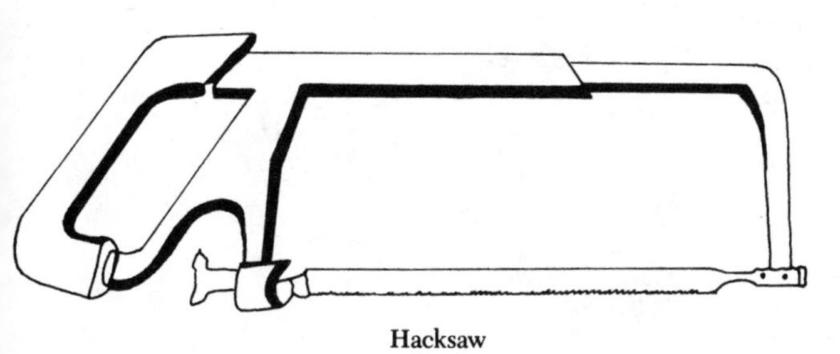

Hacksaw

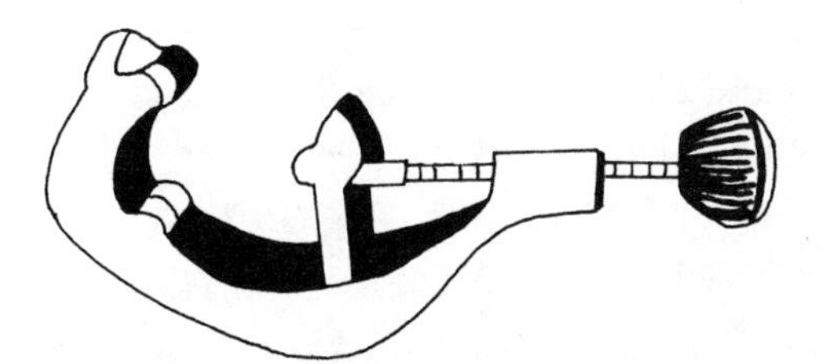

Pipe and tube cutters

Pipe and tube cutters—These are inexpensive, save a great deal of labor, and produce a straight cut. The tubing cutters have a small reamer attached to them for cleaning burrs from the severed pipe end.

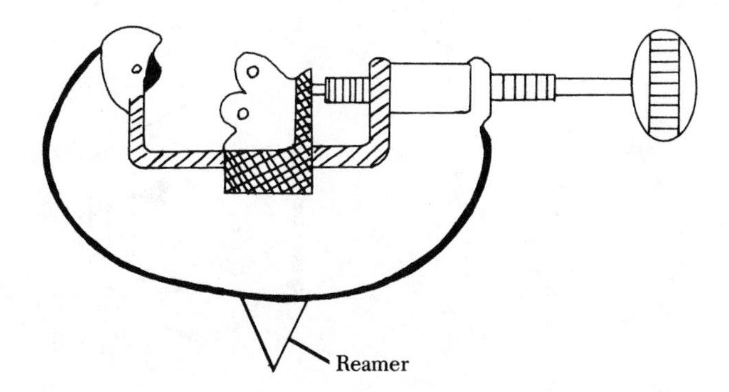

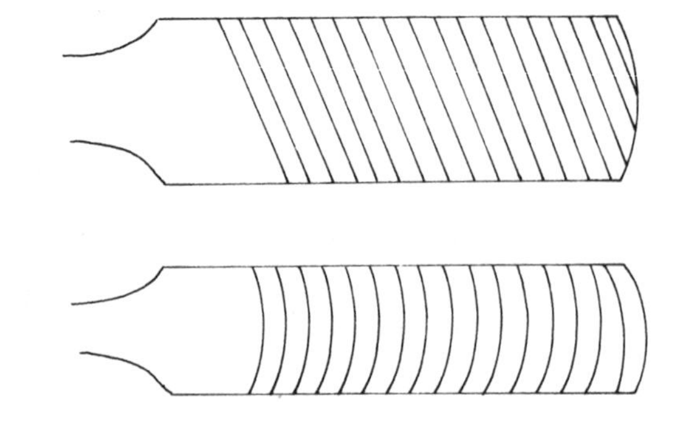

Files—Be sure to buy the type used for metal, not wood. You should own a fine, a medium, and a coarse file as well as steel wool and emery paper for removing any roughness in pipe ends to be joined with a fitting.

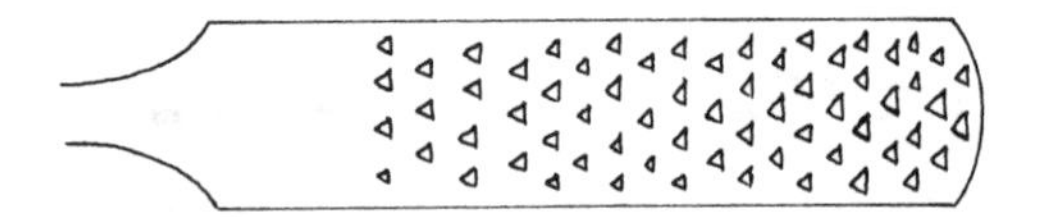

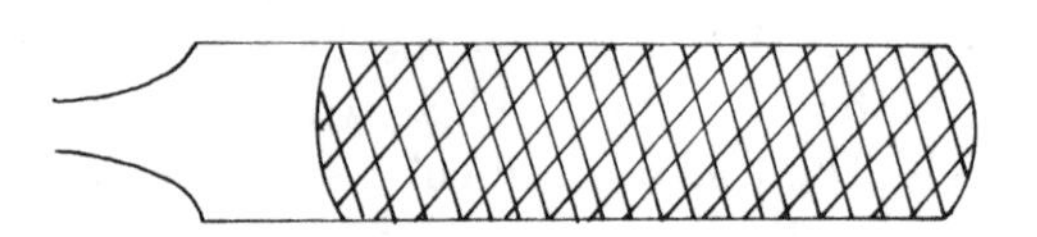

Flaring tools —Either the screw-type or the impact flaring tool will flare the ends of both copper and plastic tubing.

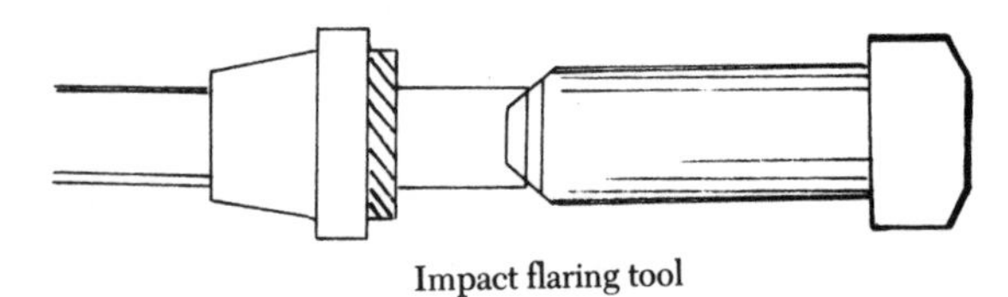

Impact flaring tool

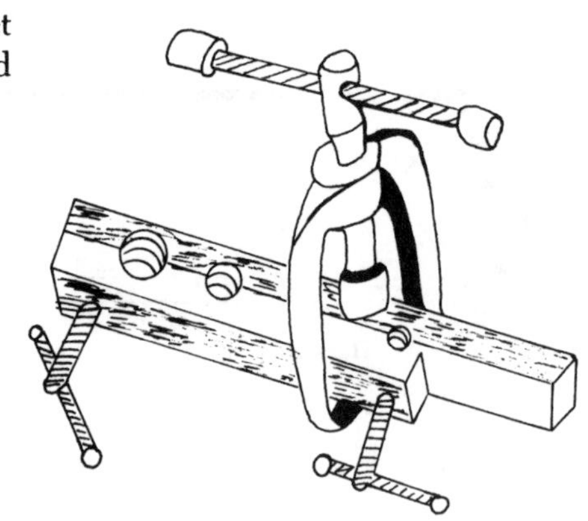

Stock and die—The same die stock will handle dies for cutting a full range of pipe diameters.

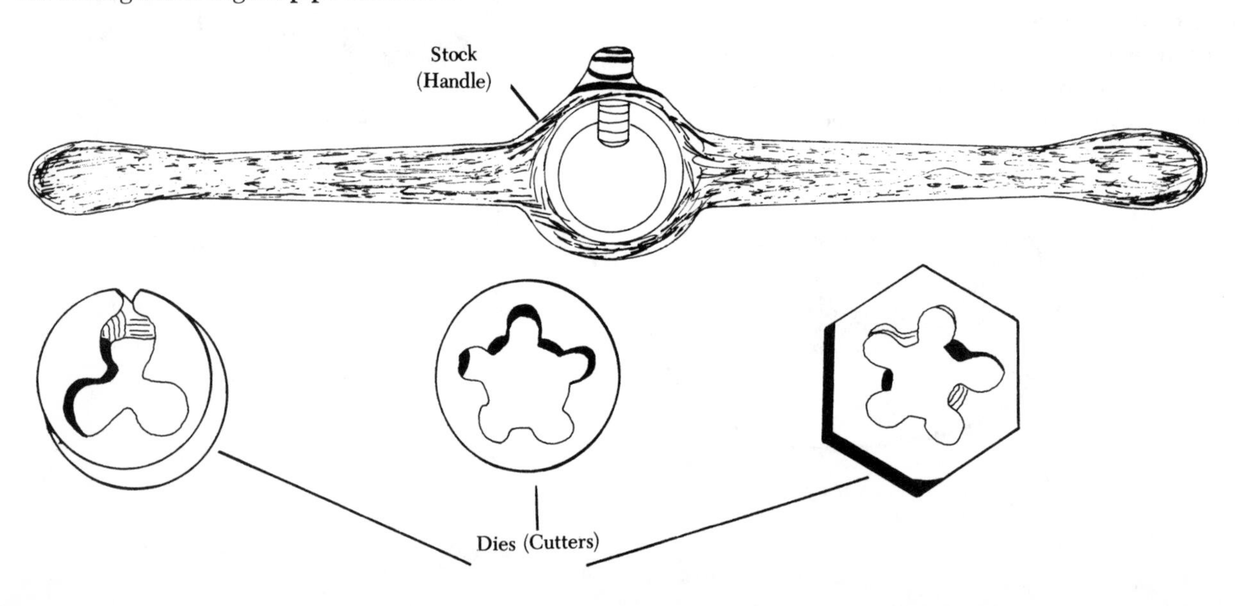

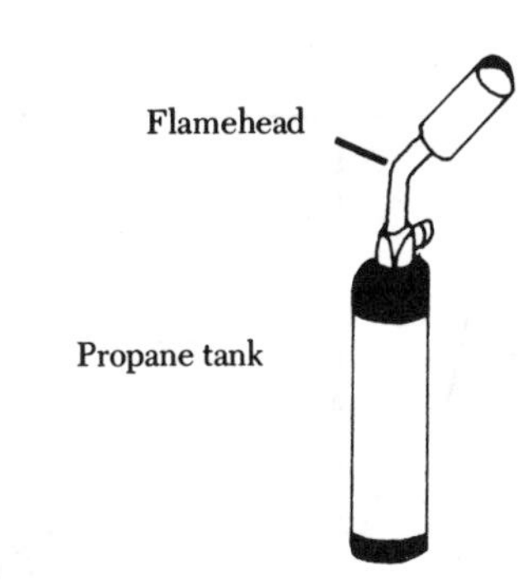

Propane torch—For about $15, you can buy a whole torch kit that includes solder, flux, two or three flameheads, and a canister of propane gas. The propane can be refilled or the canister replaced for very little cost.

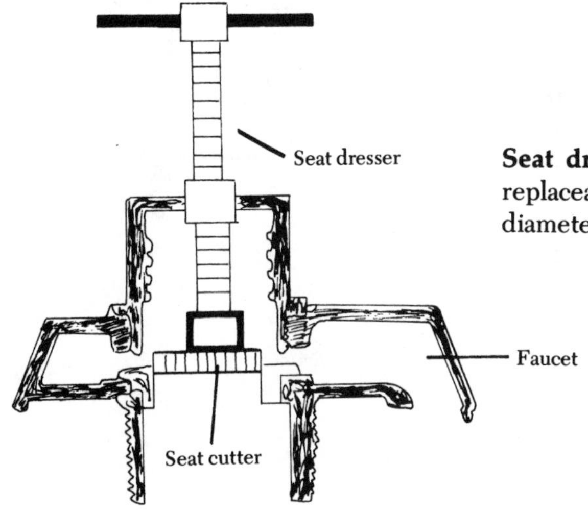

Seat dresser—These are really only screws with replaceable cutting heads for smoothing different diameter seats.

Pipe vises—A standard wood-working vise cannot grip circular pipe very well, so the important element of pipe vises is their curved jaws. But they are expensive and do not warrant the cost unless you have a lot of pipe work to do.

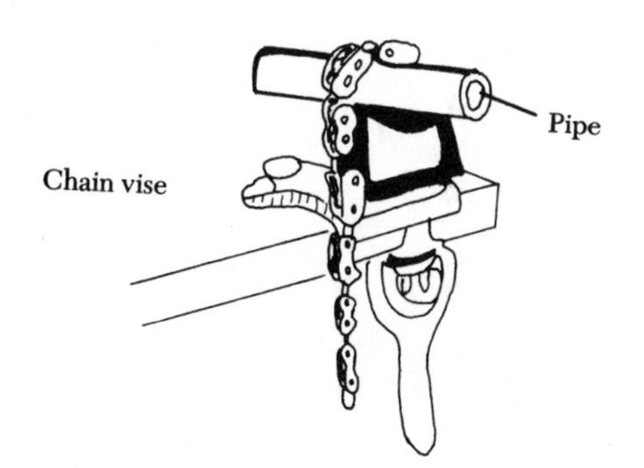

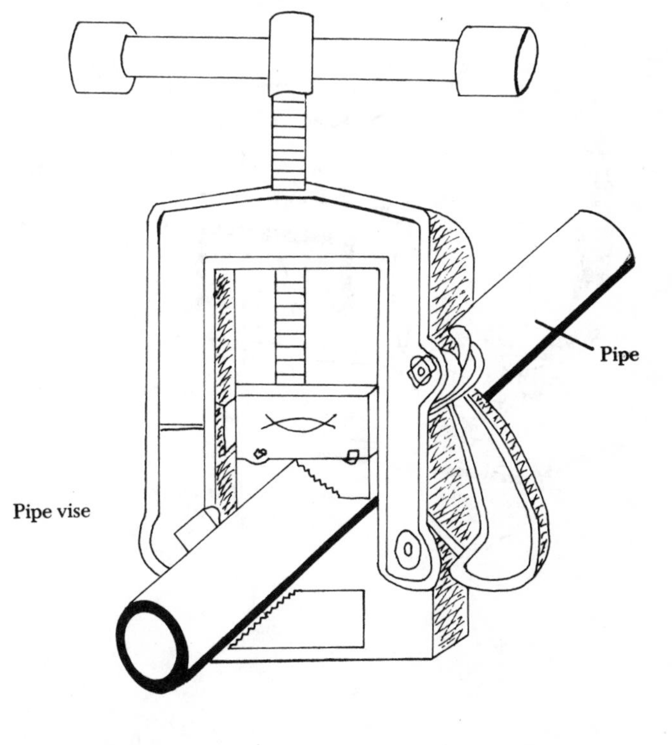

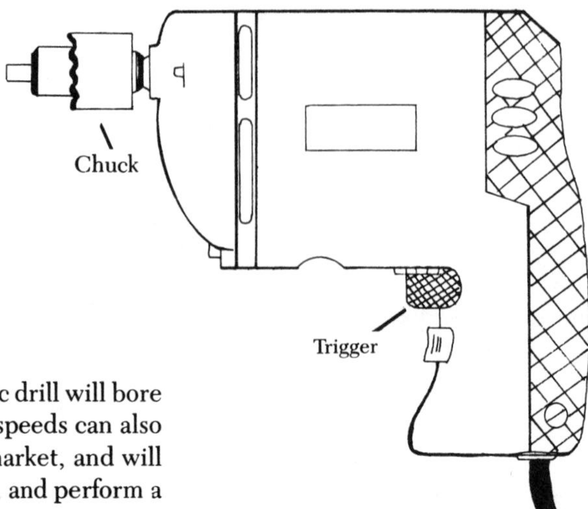

Electric drill—With high-speed bits, an electric drill will bore through any material. A 3/8"drill with variable speeds can also handle any of the accessories available on the market, and will let you drill, sand, file, pump water, saw, shape, and perform a host of other repair activities that arise in any home.

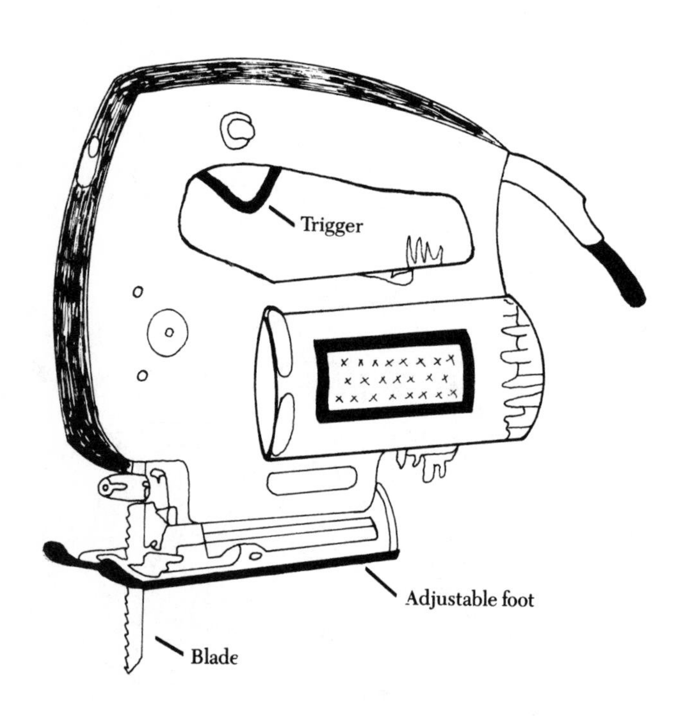

Saber saw—Inexpensive and versatile, the saber saw has a wide range of blades that will let you cut through any material from cork to cast iron. Like the electric drill, it belongs in your tool kit for making all kinds of home repairs.

INDEX